'UNTO THE HILLS'

THE HISTORY AND WILDLIFE OF
THE SOUTH DOWNS

By the same author

A Natural History of the Cuckmere Valley, The Book Guild, 1997 (Third Edition)

The Mountain of Mist – A Novel Based on a True Story, The Book Guild, 1998 (Second Edition)

The Sun Islands – A Natural History of the Isles of Scilly, The Book Guild, 1999

'UNTO THE HILLS'

THE HISTORY AND WILDLIFE OF
THE SOUTH DOWNS

Patrick Coulcher

Patrick Coulcher

The Book Guild Ltd
Sussex, England

First published in Great Britain in 2001 by
The Book Guild Ltd
25 High Street
Lewes, East Sussex
BN7 2LU

Typesetting in Times by
Acorn Bookwork, Salisbury, Wiltshire

Origination, printing and binding in Singapore under the supervision of
MRM Graphics Ltd, Winslow, Bucks

A catalogue record for this book is available from
The British Library.

ISBN 1 85776 586 9

To my ancestors who once walked these hills and to my friends who share with me the spirit of the Downs

CONTENTS

LIST OF MAPS

LIST OF COLOUR PLATES

ACKNOWLEDGEMENTS

I am indebted to many people who helped me in my research for this book. My thanks to David Compton of West Meon, Allan Brown of Brighton, Jean Davies, Janet Faulkner, Leslie Goode, Joy Preen, Andrew Wells and Marian Wood for their advice and contribution.

My thanks also to the many members of the Sussex Botanical Recording Society, particularly Frances Abraham, Peter Davys, David Lang, Janet Simes, Dennis Vinall, and to those of the Sussex Wildlife Trust, especially Robin Crane MBE for his comments on the proposed National Park for the South Downs.

I am grateful to the following for allowing me to include the stunning photographs in the book: Michael Hollings, Brian Meldrum, Dr. Paul Maurice, Melvin Smith and Dennis Vinall.

I thank particularly Mrs. V. Wootton for kindly allowing me to use a painting of her late husband Frank Wootton, OBE, as the cover for the book, and those other talented artists Andrew Dandridge, Gillian Keefe and Christopher Osborne for their wonderful sketches and paintings.

My thanks and appreciation also to all those hard working people at The Book Guild Ltd who have done so much to produce this book.

Finally thank you, Denis Healey, for doing me such a great honour again in writing the foreword.

FOREWORD

By the Rt. Honourable Lord Healey of Riddlesden, CH, MBE.

Anyone who enjoyed Patrick Coulcher's *A Natural History of the Cuckmere Valley* will be as delighted as I to find that he has now broadened his scope to cover the South Downs as a whole. For he has an exceptional ability to describe a landscape and the wild life which it shelters.

The fifty miles of downland between Eastbourne and Winchester look south to the sea and north to a Weald of comparable beauty. The chalk cliffs whick stretch from Beachy Head along the Seven Sisters to Seaford have no equal in Britain. Kipling's 'blunt, bow-headed, whale-backed downs' have inspired some of our greatest literature – the prose of Richard Jefferies and Virginia Woolf and the poetry of Tennyson, Hopkins and the so-called Georgians in the first quarter of the twentieth century.

Patrick Coulcher has a deep understanding of how landscape has been shaped by history – and prehistory. Many will share his concern that sheep and cattle are destroying the long grass which provides for so many birds, butterflies, plants and flowers – though for me the sheep are an inseparable part of the downland's beauty.

All of us will agree on the vital importantce of conservation and welcome the urgency with which he makes his case. However, the value of this book will be to enrich the life of its readers by intro- ducing them to the full beauty of an area still too little appre- ciated by those not lucky enough to live there.

Denis Healey

INTRODUCTION

Wolstonbury Hill

'I will lift up mine eyes unto the hills, from whence cometh my help.'

PSALM 121

My first sight of the South Downs was from a train travelling from London to Brighton on a bright sunny afternoon in the early summer of 1947. As the train sped southwards the line of these southern hills gradually became more evident, their outline more distinct, set against the deep blue sky and puffy white clouds so delightfully characteristic of this part of England. Little did I realise then how this long rampart-line of low chalk hills would fascinate and absorb so much of my time and give me so much pleasure in later years.

These 'blunt, bare-headed, whale-backed Downs' so beloved and described by Kipling have a huge measure of history and much to reward the naturalist in their beauty, their variety of birds, trees and flowers and other abundant wildlife.

In this book, the South Downs are defined as the line of hills stretching some 110 kilometres from Beachy Head in East Sussex, to St Catherine's Hill near Winchester in Hampshire. I am very aware that there are many interesting places close to the Downs themselves, in the Weald, the river valleys, the coast and the coastal plain. I have decided therefore to include many of these in this book, so you may find pleasure in the description of places such as Tide Mills (Newhaven), Amberley Wild Brooks and Pagham Harbour.

Just as before, in my books on the River Cuckmere and the Isles of Scilly, I will keep the structure of the book simple, describing first the geology and history of the South Downs, and then outlining the natural history of defined and separate areas. Finally, in the postscript, I touch upon contemporary problems associated inevitably with conservation and how our heritage can be protected and safeguarded.

This book is not intended to contain a full and precise description of every natural species along the Downs; to do so would require an encyclopedia, nor is it intended for the professional naturalist, although he or she might find something of interest in it. Instead this book describes species, (particularly birds, plants and butterflies), which have beauty and interest, or are peculiar to the area. As such it will appeal to those seeking knowledge about the natural world, and to those who just want to relax, perhaps at home, and read about a lovely part of our British countryside.

I am very conscious of the fact that many books on the South Downs have been published recently, all with their own style and content. This one is intended to complement them and especially to please the reader in its descriptions of the history and atmosphere of the Downs and interesting places on their periphery.

A book such as the *AA Book of the British Countryside* will be useful for identifying the butterflies and other insects mentioned in the text. Binoculars and a book such as *The Field Guide to the Birds of Britain and Europe* are essential to learn about and recognise birds. There are many good books to choose from to learn about plants; perhaps the best for the beginner is W. Keble Martin's *The Concise British Flora in Colour*. Do buy a good hand lens (10 magnification) to examine the detailed structure of plants, and you will not be disappointed by the sheer beauty revealed in their make-up.

The maps in this book are designed to highlight most, but not all, of the places mentioned in the text. For simplicity they are schematic and are not designed to show every wood, river, path or road. Additional maps, such as the 1:50000 Ordnance Survey Landranger series, numbers 185 to 199, are essential and these will show details of public footpaths and many features not explained in the book. Armed with such few items, the reader will be well equipped to plan and learn from an excursion to any of the areas I am going to describe.

Please remember at all times to fasten gates, control dogs, guard against fire, leave no litter, keep to public footpaths and protect the plants, trees and wildlife.

I hope that after reading this book the reader will be left with an enquiring mind and a firm resolve to do something positive to help preserve what is left of the South Downs, already ravaged enough by the exploits of man.

WARNING: Medicinal properties of some plants and fungi are given in this book. In no circumstances should readers experiment or try these out themselves without sound professional advice.

Beachy Head to St. Catherine's Hill

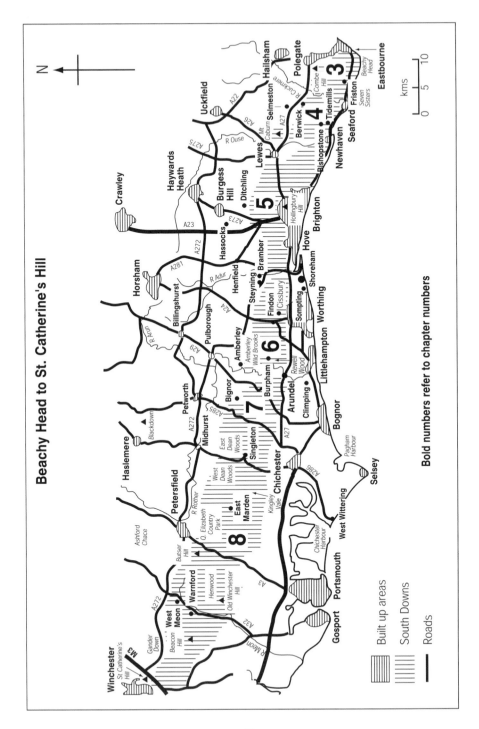

Bold numbers refer to chapter numbers

xviii

1

A GEOLOGICAL HISTORY

The North Escarpment looking west from Devil's Dyke

Some 120 million years ago at the beginning of the Cretaceous Period and well before Homo Sapiens appeared on the planet, the whole of south-east England was covered by a shallow freshwater lake into which flowed many rivers. These deposited large quantities of clay and sand into the lake whose shallow waters were the home of enormous reptiles such as the iguanodon and the plesiosaurus, the remains of which have been found in various places in Sussex. The movement of the earth's crust caused these sands and clays gradually to sink, and the lake became deeper until eventually the clays themselves attained a depth of nearly 200 metres, forming the Wealden clay we know today. This surrounds Horsham and extends in a belt south eastward to Hailsham and the Pevensey Levels. The lands of the south-east, which of course were still part of the continent, continued to sink and eventually the sea broke in.

So some 100 million years ago the sea itself began to lay down the structure of the South Downs, for it was the sea that then deposited massive layers of sand we now call the lower greensand, in places 100 metres thick, to form low ridges, close by and parallel to, the northern escarpment of the Downs. These sand ridges were important to the hunters of the Mesolithic period, as we shall see later.

The land continued to sink, and just as deep-sea oozes form today on the floor of the Atlantic Ocean, so then did layers of clay form one upon another. This gault clay formed some 100 metres thick, to be followed later by another bed of sand – the upper greensand which can be seen so clearly as a muddy-green stratified layer of hard rock off Beachy Head and Eastbourne. As the subsidence of the land continued so the sea became deeper and deeper, forming an ocean of immense depth. At this time, 100 to 75 million years ago, the world was enjoying a tropical climate, and in the warm oceans lived innumerable microscopic organisms – Foraminifera. These ancient forms of plankton were rich in calcite (calcium carbonate) and when they died their bodies sank to the bottom to form a thick white calcareous ooze. This layer of ooze formed the chalk we see today. Interestingly, amongst the debris of the recent fall of chalk at Beachy Head can be found huge pieces of chalk in the form of crystal called calcite. Some are imbedded in huge chalk boulders and are beautiful to see. Anyone looking at the great chalk cliffs will not fail to see how pure this layer of chalk appears to be, and it must be concluded that the seas at that time must have consisted of clear water unsullied by mud or sand brought from the land. It has been suggested that the land area surrounding the sea then consisted of desert where little rain fell and no rivers or streams flowed. This would explain the abnormal purity of the chalk, free as it is of sediment.

Nobody who views the great chalk cliffs of the Seven Sisters from the beach can fail to notice the dark bands of flint at almost regular intervals of a metre or so, up the cliff face. Flint, like the chalk, has a biological origin and was formed from minute animals called radiolarians that floated with the plankton of those ancient seas. These animals together with siliceous sponges that lived on the seabed had skeletons which after decay formed into lumps of crystalline silica or flint nodules. Nobody knows for sure why these nodules formed bands in the chalk but I believe that the radiolarians and sponges required a certain sea temperature to bloom and flourish, and as the earth's axis tilted back and forth over tens of thousands of years (part of the Milankovitch effect), the sea temperature rose and fell correspondingly. Thus the flint nodules formed in bands on the seabed according to the temperature fluctuations over the millennia. As we will see later, flint played an important part as a raw material for use by neolithic people.

The chalk which formed during the Cretaceous period extended from France and Belgium to north-eastern Ireland, and from Dorset to the Yorkshire Wolds and Denmark. After this period some 65 million years ago, the seas became shallower and the earth entered what we now term the Tertiary period. It was at the beginning of this period that the ocean floor began to rise and as the water grew shallower so the formation of chalk ceased, giving place to deposits of clays, loams and pebbles. These deposits are known by geologists today as the Reading and Woolwich beds and soon these too were covered by later deposits such as the deep layer of London clay. One interesting deposit on top of the London clay was that called Bracklesham sand now found only near Selsey Bill. Fossils recovered from this layer include crocodiles, turtles, aquatic serpents (one over 6 metres long), shells, corals and the fruit of palm trees.

The Tertiary period was the time when the present-day division of land and water took shape, as the molten inner core of the earth contracted, causing the outer crust to buckle into great folds. In this way huge mountains were formed and in our own area of the continent the strata crumpled and overturned on a gigantic scale, leaving us with the Alps. The ripples of this great earth movement or 'Alpine Storm' as it is known, affected Southern England; the thick bed of chalk together with its associated strata of sands and clays was gradually pushed up into an immense elongated dome over an area stretching from Hampshire to Calais in France, and covering the whole of what we now call The Weald. This dome at its highest point, probably over Mayfield in East Sussex, achieved an elevation of over 1000 metres above the surrounding area. As the buckling of the underlying strata continued so great folds began to form, giving rise eventually to the Thames valley in the north and what was to become the English Channel in the south. In between, the chalk at the top of the dome began to fissure and disintegrate and the falling rain soon carried off these fractured pieces of chalk from the summit. This denudation of the dome continued for millions of years, and of the streams that formed on the slopes, those that flowed north joined with others to form the Thames, while those that flowed south fell into the fold that some 7,000 years ago, eventually became the English Channel and emptied westwards into the Atlantic. It is these south flowing streams that now neatly divide the Sussex Downs on a north-south axis and provide a suitable way of dividing this book into its various chapters.

Nearly a million years ago the last great episode in the geographical evolution of this country began; the ice age. It is very difficult to reconstruct the exact sequence of events of the ice age but patient field and laboratory observations over many years have elicited certain facts. Some 600,000 years ago the northern latitudes of the planet became much colder and a great ice-sheet covered Scandinavia, the North Sea and Britain at least as far as the Thames. There followed intervals of warmth called the interglacial periods and these alternated with the colder periods. There were a total of four glacial periods with cold phases lasting some 100,000 years with interglacial periods lasting 50,000–200,000 years. The last glacial period ended about 8,000 BC and since we must regard ourselves as being placed firmly in an interval between recurrent cold periods, we may reasonably expect further glaciations to occur. Of course the present phenomenon of the greenhouse effect could temporarily delay them but eventually the glaciers will return.

What caused the recurrent glaciations? Almost certainly they are due to the periodic tilting of the earth on its axis back and forth over tens of thousands of years, caused by the mutual attraction of the planets with the earth as it orbits the sun. This tilting results in changes in the amount of the sun's radiation on the earth.

During the glaciations the sea rose as ice melted during the warmer interglacial periods, and then fell as its waters were stored in the form of snow and ice on land and when it was cold. This geographical phenomenon formed raised beaches which can clearly be seen as pebbles of old seashores at a height of between 20 and 45 metres above the present sea level particularly in quarries and gravel pits between Arundel and Goodwood. At Black Rock just to the east of Brighton behind the marina, another raised beach at about 5 metres above present sea level can be seen in the lower part of the cliff. Anyone passing along the undercliff walk towards Rottingdean can easily examine this raised beach in its form of rounded beach pebbles lightly cemented together. Above it the rest of the cliff consists of a deposit of chalk rubble; a mixture of angular pieces of chalk, flint chips and large lumps of sandstone. This collection of rocks is known as coombe deposit and lies spread out as a thin sheet across a large part of the coastal plain of Sussex. It is also found in the dry downland valleys or coombes, particularly at their lower ends; hence their name. If you visit Birling Gap just to the west of Eastbourne you will see a good example of this combe deposit in the low cliff as you walk down the steps to the beach.

So we end this brief analysis of the geological history of the Downs and their environs. Of necessity this has been a short and simple description of how our chalk downs were formed; much, much more could have been written, but to do so perhaps would have been of interest only to the geologist and would have been tiresome to those of you whose main interest is in the plant, insect and animal life of the South Downs. However, by touching on the subject I hope your interest in the structure and scenery of our beautiful countryside will have been stimulated, for as we explore the ancient downland valleys, the sea coast and places like Mount Caburn and the Ouse valley, you will see that their natural history is much influenced by the underlying geology.

Sit one day on top of Haven Brow, the first of the Seven Sisters overlooking Cuckmere Haven, where the Cuckmere flows into the English Channel. This is a place unique in Southern England where a river flows into the sea unaccompanied by a port complex, roads and cars, caravan sites and other trappings of the human race; a place still comparatively remote and peaceful where the silence is only broken by the call of a seagull or the shrill piping of a flock of *oystercatchers*. It is early morning in midsummer, and the heat of the sun begins its work of warming the air, causing it to rise and then condense into puffy white clouds that begin to form like bundles of cotton wool over the land. To the west the nearer cliffs are coloured brown with the pebbles, sand and clay swept down by ancient ice fields and overlying the older chalk; beyond, the great bastion of Seaford Head stands out starkly against the distant horizon. How long will its majestic presence last against the ever-pounding waves gradually eating away at its soft protuberance? To the north, see the broad sweep of turf-covered downland undulating in smooth waves, and view the distant dark hills of the Weald which beckon you to explore their quiet woods and green fields. Beneath you to the east are the sheer white cliffs of chalk shimmering and glistening in the bright light of the sun. Feel the peace and solitude of this place, and let the limitless sky and the gentle lapping of the waves below lull your senses and empty your mind of the present; dream of times long ago when the chalk itself was being laid down on the floor of warm seas, and when the great Alpine Storm uplifted mountains, to be followed by that great ice age which moulded the landscape all around.

Abruptly you awake to the dull roar of a modern jet plane high up above, its white vapour trails carving a path across the broad expanse of blue sky. Your mind quickly turns to man himself, his evolution, and how he might continue to alter the earth's environment and change the very nature of the hills where you are lying.

Beachy Head

Lewes Castle

2

SOUTH DOWNS PEOPLE – A HISTORY

The Seven Sisters

'Trackway and camp and city lost,
Salt marsh where now is corn,
Old wars, old peace, old arts that cease,
And so was England born.'

RUDYARD KIPLING

Palaeolithic Age – Before 7,000 BC

When did the first humans touch upon the landscape of the South Downs? Nobody really knows but recent discoveries at Boxgrove 10 kilometres west of Arundel, indicate that man used the site some 500,000 years ago. Boxgrove Man, as he is now known, represents Europe's oldest human remains. With these remains were found several hundred hand axes and bones of carefully butchered animals indicating that this place was used to corner and kill animals such as deer and wild horses. Clearly, man has been around for much longer than previously supposed and during the interglacial periods our ancestors must have arrived and retreated with the climate.

There is evidence that more continuous occupation of the Downs began some 12,000 years ago when people of the Old Stone Age (or palaeolithic) began to inhabit the area. Then the great ice sheets were retreating and the climate was slowly warming, and these palaeolithic people migrated from the continent which was at that time still joined to Britain. In this dim and distant age our planet was sparsely inhabited by these people who generally lived by the sea or in river valleys; they were hunters and food gatherers with a nomadic or semi-nomadic existence, moving on when local food supplies ran out or to follow migrating animals. Although they left comparatively few traces in Southern England where their population could scarcely have exceeded a hundred or so, a few of their pear-shaped hand axes have been recovered in Sussex. Our understanding of how these people lived can only be conjecture but their rude tools of flint and bone indicate that they hunted elephants, hippopotamuses, bears, horses and deer. Fishing, too, was important to them and their artefacts have been found in river valleys and on sites of ancient raised beaches. Little is known of their dwelling places; perhaps holes in the ground, in trees, or in caves of which there must have been few in Southern England. The South Downs themselves would have presented a bleak, tundra-like appearance, with little vegetation and on at least four occasions the weather became arctic with intervening periods of warmth. Man would have retreated southwards on the approach of ice and returned when it receded. Our knowledge of palaeolithic man is sparse and can be summed up from the writings of the Roman poet, Lucretius, 'in ancient times a man's weapons were his hands, his nails and his teeth, with stones and branches broken from the trees, and also fire, when this had been discovered.'

We may know little about these people but their story absorbs by far the greatest period in human history, embracing hundreds of thousands of years.

Mesolithic Age 7000–3000 BC

The first of the mesolithic, or Middle Stone Age people, probably walked into Southern Britain from the continent before the rising sea of a warmer climate finally cut us off from the continent in about 5,000 BC. With the passing of the ice ages the ice sheets retreated northwards to be replaced with, at first, cold-loving trees such as *birch* and *willow*, then trees that required a warmer climate such as the *oak*, *ash*, *hazel* and *Scots pine*. Thus forests grew where once there was just tundra and steppe vegetation.

Mesolithic people knew little or nothing of agriculture or the domestication of animals and like their predecessors still hunted game, but with improved equipment. They learned to fashion flint into weapons in the form of microliths (from the Greek, meaning 'small stones') used on spears and arrows, and made axes shaped from blocks of flint stone. Mesolithic people not only hunted and fished but also gathered the fruits of the vegetation all around, such as hazelnuts and berries of all kinds. These hunter-gatherers would have made lengthy forays into the heavily-forested Weald country using well-defined trackways; but at the end of the summer they would have returned for the winter to their river fishing grounds on well drained soils near to the Downs. Evidence of their preference for dry sandy rides has been found at Hassocks and Selmeston. Excavations behind the church of Selmeston in 1933 revealed mesolithic dwelling pits and over 6,400 worked flints. Such homes were probably roofed with hazel boughs and thatched with heather, and within was a fire on which food was cooked and water was heated in skins from hot stones. Here, they shaped their flints and made axes, bows and arrows, and fish spears tipped and barbed with sharp microliths; but a change was on its way as cold dry north-east winds gave way to warm wet south-westerlies and *oak* trees appeared amongst the *hazels*. Such a change accompanied the dawn of civilisation, with the advent of neolithic or New Stone Age man, who was not merely a food-gatherer but also a food-producer.

Neolithic Age 3000–1500 BC

The neolithic people were notable and quite distinct because they had the knowledge of the components of civilisation, such as the growing of corn, the herding and domestication of animals such as the ox, sheep and pig, the making of pottery and the grinding and polishing of flint tools and weapons. They were probably the first to make clearings on the heavily wooded Downs in order to establish a permanent camp or village site. Around these sites they made enclosures surrounded by earth ramparts in which animals were herded in autumn for the slaughtering of the weakest that could not be maintained through the winter. The remains of such neolithic enclosures can be seen at the Trundle above Singleton, Barkhale Down above Bignor, Whitehawk at Brighton and Combe Hill above Jevington.

With primitive hoes and sickles made from bone and flint these early people grew wheat and barley in small clearings using skills brought with them from mainland Europe. They also fashioned pottery and extended the already established flint mines on the

Downs (some have been carbon dated to before 4000 BC). Flint, as we have already described, is found in bands within the chalk; neolithic people discovered that those flint nodules found exposed in the open for any length of time were tougher and less workable than those excavated from the interior of the chalk. They therefore excavated and built flint mines at suitable places on the chalk escarpment, using deer antlers, and shovels made from the shoulder blade of an ox, deer or pig. Flint excavations can be seen at such places as Cissbury (north of Worthing), Harrow Hill (north-west of Findon), Stoke Downs (north of Chichester) and Windover Hill (near Wilmington).

The Downs as we know them today were originally cleared and moulded by these early farmers, bringing with them the dawn of civilisation into the south of England. As Rudyard Kipling so aptly puts it in *Puck of Pook's Hill,*

> And see you marks that show and fade,
> Like shadows on the Downs?
> O they are the lines the Flint men made,
> To guard their wondrous towns.

Bronze Age 1500–500 BC

The transition from the neolithic period to the Bronze Age was extremely gradual. Bronze is an alloy of copper and tin and had many advantages over copper alone, being tougher and more easily melted. Bronze was thought to have been used as long ago as 2000 BC in the eastern Mediterranean and was brought to England by a new race of people who had different physical features to the neolithic natives. On average they were several inches taller, were more robust with broad heads and rugged features, contrasting with the neat narrow faces and longer heads of neolithic people. This new race came from the lower Rhine districts of Germany and they brought with them, as well as bronze, distinctive pottery in the shape of beakers. These finely-made drinking cups of pinkish red clay had thin walls and were decorated with impressed patterns mainly of a horizontal nature. These vessels gave rise to the commonly held name of these people, 'Beaker Folk'.

Little is known of how they lived, but in 1949–53 a site at Itford Hill near Newhaven was excavated and revealed thirteen circular huts, each with a central post and smaller posts around the perimeter. Numerous pieces of pottery were found on the site and an enclosure system for animals was also evident. Nearby were small irregular fields marked by earth banks. These were formed by the continuous ploughing of one section of the downland contour in order to get some depth of soil to plant crops. These 'lynchets' as they are called, derive their name from a Saxon word meaning 'little hill'.

Bronze Age people buried their dead in circular burial mounds which covered their cremated remains, contained in pottery vessels. These round 'bell' or 'bowl' barrows (depending on their shape) differed from neolithic burial sites where their chieftains were buried in long or oval-shaped mounds flanked by ditches.

So with the development of small hut villages, primitive ploughing and complicated field systems, the clearance of the Downs proceeded apace.

Iron Age 500 BC–43 AD

Around 500 BC new invaders from mainland Europe, the Celts, came to the shores of Sussex. They brought with them a new material; iron. Iron, which was immensely strong, had been worked in the eastern Mediterranean as long ago as 2000 BC, and gradually its use spread westwards to Western Europe. The Iron Age Celts were artistic, talented and warlike, and they built massive ramparts on hilltops, sometimes obliterating neolithic mounds already there. Good examples of their sites can be seen at Park Brow (near Findon), the Trundle (near Goodwood), New Barn Down, near Harrow Hill (Worthing) and Mount Caburn (Lewes). A massive hilltop fort was also built at Seaford Head but this has mostly fallen into the sea. The Celts gradually took over from the Bronze Age people; they made great improvements in farming techniques and established their own tribal system, dividing the chalklands into regions. Celtic ironsmiths improved on the previous wooden plough with one tipped with an iron shoe, and they were able to extend and enlarge the lynchets and construct more complex field systems in which to grow their barley, wheat, rye and oats.

The original Celts came first from Hallstatt, a site in Austria, but later in about 250 BC they were followed by a superior group from the Marne area in Northern France. These later Marnian Celts brought with them the horse-drawn chariots which were to prove so formidable against the Roman legions later on. There is little doubt that the later Celts made forays into the Weald and established the first iron workings there, using the extensive deposits of the metal in the area. Certainly there is some evidence of this in the Iron Age pottery found in some of the early slag heaps. Two fortified camps have been found at Pipers Cross at Kirdford and a hill at Saxonbury, near Frant; these were probably mining camps protected not only from human interference but also from *wolves* which were numerous at that time.

During this Celtic period the Downs were beginning to look a little as they are today, with areas of ploughed land and cultivated crops, but unlike the present, in those days people lived on the top of the hills. The climate was colder and wetter than it was in the Bronze and neolithic ages and the water table was higher with many springs emanating from the lower slopes, giving a good supply of water to the inhabitants.

The Celts, like the people before them, would have helped themselves to the nutritious plants that nature provided. So they would have gathered plants like *seaweed* and *sea kale* from the beaches and weeds such as *chickweed* from the fields. *Hazelnuts* and *blackberries* would have been plentiful and they may have used plants such as the *field fleawort* for medicinal purposes. It is interesting that field fleawort seems to grow most commonly near Iron Age sites and trackways. It is not known in France and could have been introduced by the Hallstatt Celts from Austria where it does grow.

When times were good and food was plentiful, perhaps our Celtic ancestor had some time to stand and stare at what he had created out of the downland wilderness of scrub and woodland. Maybe like us he would have seen the sunset in a blaze of glory on the skyline of the distant sea whose shoreline then was perhaps quite a number of kilometres further away than today. Perhaps too, he would have contemplated life as we sometimes do now, with its complex human relationships; where love can turn to hatred almost as quickly as an autumn leaf falls to the ground on a breath of wind, and anger is appeased by the quiet counselling of a friend. Did he then, as we do now, look into the future to see what the world was all about and why he was put upon it in the first place? We, of course, have the occasion and space to think on these things, but he had little time to stand and stare and was truly busy in just keeping himself alive with enough food, water, warmth and shelter.

Of course each of the different ages we have discussed so far did not suddenly begin when the previous one ended, but fused and overlapped; the Iron Age merged with the Bronze Age just as the Bronze Age blended and intermingled with the neolithic and the neolithic with the mesolithic.

The Roman Period AD 40–410

The main Roman invasion which occurred in AD 43 resulted in a permanent occupation of nearly 400 years. As conquerors, the Romans were brutal and unrelenting but once the hostility of conquest was over they were magnanimous and displayed an extraordinary capacity for obtaining the loyalty of their new subjects. The invaders themselves were a racially mixed bunch of peoples coming not only from Italy and Rome but also from the then known world; France, Germany and Spain and even Arabia and Africa. These people integrated well with the Britons and certainly in Sussex little resistance to the new aggressors was offered. Perhaps only at Mount Caburn (the name derives from the Celtic term 'Caer Bryn' meaning 'Fortress on the hill') were there signs of any significant opposition. The occupants of this settlement, who were descendants of those Iron Age people who had first occupied it in about 300 BC, decided to resist the Romans. They reinforced the existing small rampart by building a huge outer one over 7 metres high on its northern side where the flat terrain made the settlement more vulnerable to attack. These preparations were in vain as the Romans overcame the defences at Caburn and the site was left abandoned. Evidence has shown that the timber work of the gateway was set on fire and nearby the remains of two Roman scabbards were found.

The Romans brought peace and prosperity to England; they built roads and comfortable villas, particularly in West Sussex, where the remains of their elegant villas, fitted out with mosaic floors and central heating, are thickly scattered around the area of Chichester, itself a city of great importance in the Roman period. The Romans enlarged the iron industry of the Weald and they improved the efficiency of farming on large estates which were worked by slaves or serfs. It is not surprising that the British, so used to feuding and conflict, soon adopted the 'Roman ways' and easily integrated with the invader to form a new people, the Romano-British.

The cultured and colonial style created by the Romans lasted until early in the fifth century, when hordes of the barbarians from the east attacked and ransacked Rome, and the Romans left Britain to protect their homeland. The Romans had brought much to Britain, an ordered and measured way of life and a system of government that gave to the people a sense of belonging. They brought with them too a legacy of medicinal plants, *alexanders*, *betony*, *white bryony* and *hedge mustard*, and reintroduced others such as the *common mallow*, *tormentil* and *hedge woundwort* which once flourished in the ice age. They also used for medicinal purposes such plants as *henbane* and *common fumitory* which were introduced in the Iron Age.

The *walnut* and *sweet chestnut* trees were brought into Britain by the Romans; both grew in Italy as long ago as 100 BC and their fruits were widely used later as a nutritious supplement to their diet. In Sussex the sweet or Spanish chestnut is common on the sandy soils of the Weald, and the walnut though less common because of its susceptibility to frost damage, can be found near old farmhouses.

So what did the Roman centurion think of the land he left behind? He must have looked across the tree-covered hills of West Sussex and in the east to the rounded smooth shapes of Firle Beacon and Mount Caburn and taken pleasure in the peace that was all around; he would have gazed and taken pride in the long straight roads his masterful engineers had cut across the wilderness of the wooded Weald. As he boarded the great timbered vessels that bore him back to Italy he would have had more than a twinge of regret on leaving such a place as the green and pleasant land of England.

But change was to take place yet again as the survivors of the Roman civilisation, the Romano-British, were left to fend for themselves against new invaders from the Continent.

The Anglo-Saxon Period AD 410–1066

It was a warming period for the earth and the great northern glaciers were melting and the seas were rising. The low-lying lands of Northern Europe were reducing in size and their peoples, the Anglo-Saxons, were hungry for land. So in the third and fourth centuries AD they came in small boats from an area just south of Denmark to raid and pillage the relatively sparsely inhabited areas of Southern England. These new conquerors, who gave Sussex its name, 'Suth-Seaxe', were most successful, and after the initial battles and skirmishes with the indigenous Romano-British people they succeeded in subduing the people and restoring a measured peace and stability to the country. Judging from the evidence of place names and agricultural practices, the Britons in Sussex at least, worn down by centuries of raiding, were virtually extermi-nated. Neither the Roman names Regnum nor Anderida (Pevensey) have survived, only two or three river names of Celtic origin remain, and as far as agriculture is concerned the Saxons introduced the English open-field system which contrasted strongly with the Celtic enclosed system established mainly on the open downland. The Saxons brought with them a heavy plough and because they were used to flat, low-lying lands they concentrated on cultivating the clay soils of the lowland Weald which they cleared of wood and scrub. Here, chiefly just under the base of the Sussex hills, they built manor houses that have now become our modern villages, whose names generally terminate with -ing or -ham. The Downs themselves which had become drier and which the Saxons avoided, were left in splendid solitude. Their bare shoulders of wild grasses and flowers were a haven for the *skylark*, and *meadow pipit* and their soft woodland glades came alive with the call of the *rook* and the *robin*.

The Saxon people were gradually converted from paganism to Christianity, mainly by the influence of St Wilfrid. St Wilfrid was the Bishop of York but was driven out from his northern diocese by the jealousies and persecution of Saxon kings. In about 680 AD he found himself in Sussex, a place heavily wooded and cut off from the rest of the country by a natural rampart, the Weald. It was St Wilfrid who once preached Christ's gospel at a time when the crops all around had failed from drought, and famine was rife amongst the people; as he spoke a gentle healing rain descended. This miracle softened the attitude of the hardened heathens and St Wilfrid gradually gained the respect and affection of them all. He showed them how to exploit the riches of the sea with nets and he built a monastery on the peninsula of Selsey, literally the Island of Seals, near Bosham. This land had been given to St Wilfrid by the recently consecrated Saxon King Ethelwatch and from this base St Wilfrid became the master of all Sussex men. He built a cathedral at Selsey but for centuries now this has been under the sea. Today, men whose ancestors he had first taught their skills, catch fish above the cathedral's ruins.

St Wilfrid became the first Bishop of Sussex. He baptised and gave freedom to all the slaves of his Selsey lands. He went on to convert the cruel King Ceadwalla who in AD 685 had slain in battle his patron, King Ethelwatch. So St Wilfrid has a very special place in the history of Sussex; with the advance of Christianity came the building of Saxon churches with their distinctive arches and small defensive windows. Notable Saxon churches can be seen at Sompting and Bishopstone.

In the late Saxon period Viking marauders from Scandinavia had subdued the North of England, but Sussex and Hampshire were little affected by their raids. Fortified settlements or burghs (the forerunner of our boroughs) were built to counter the Vikings. In 894 the men of Chichester killed many of the enemy and captured most of their ships. In 1035 a Saxon prince, Edward the Confessor, was elected King; he owned a good deal of land in Sussex including three manors at Exceat in the Cuckmere Valley, and also Berwick, Claverham and part of Frog Firle, all on the western side of the river.

So at the end of the first millennium AD in Anglo-Saxon England, the foundations of much of our present culture and way of life were laid, and although rival chieftains and princes fought amongst themselves, the seeds of a prosperous land were being sown; but with the death of Edward the Confessor all this was to change.

The Norman Invasion

The Norman invasion, with the main fleet landing at Pevensey, is well documented, and resulted in William, Duke of Normandy, being crowned King of England on Christmas Day 1066. The security of the Sussex ports from which the Conqueror and his immediate successors sailed when they visited their Norman duchy, were important to William and he quickly took action to secure them. He divided Sussex into six areas or 'rapes' (derived from an Old English word 'rap', meaning rope, indicating a roped-in open-air court of Saxon origin). Each rape was entrusted to a powerful and loyal baron of the Conqueror, and each of these lords built himself a strong castle from which to rule his dependants in a feudal hierarchy. Hastings was given to the Count of Eu, Pevensey to the Conqueror's half-brother Robert of Mortain, Lewes to William de Warenne, Bramber to William de Braose, while the rapes of Chichester and Arundel were at first held by Roger de Montgomery, a man of strong character and personality. So a feudal system which was first established in Southern England then spread to the rest of the country. All land was held by the King through his tenants-in-chief, the lords of the rapes, who exercised authority from their castle strongholds. Lesser lords built manor houses and everywhere the new masters rebuilt towns, villages and churches in the bold and sturdy Norman style. Good examples of Norman churches with their rounded arches, stout walls, massive pillars and striking designs can be seen at Steyning, Bramber, Burpham and Climping. The Norman influence continued for centuries and eventually they were integrated into the English nation, and their speech became blended into the English language.

As for the Downs themselves, these had been progressively cleared and in the eastern sector where there were good areas of water meadows in the Ouse and Cuckmere Valleys, *sheep* were extensively reared for their wool (it was not until much later in the 18th century that the emphasis switched to breeding sheep for their meat). Good water-meadows were important to sheep-rearing. Downland grass becomes thin and sparse in winter and for the spring lambing season the sheep had to be driven down to the water-meadows where the grass was thicker and more abundant. Since prehistoric times sheep had been used to fertilise the thin chalky soils of the open downland. Put into folds overnight on arable ground, they trod their dung in, allowing good crops of wheat and barley to be produced. This close relationship between sheep and arable farming on the chalk Downs continued until the advent of artificial fertilisers in the mid-19th century.

So for hundreds of years the Downs remained a quiet haven of peace and restfulness, undisturbed except for the soft rustle of the wind through grass, the song of birds, and the occasional call of the shepherd to his flock of wandering sheep, their bells tinkling through the early mist of a summer dawn.

After the Normans

The Normans were the last successful invaders of England, but too often in the centuries that followed England was at war with France, Germany, Holland or Spain. The Hundred Years' War of the fourteenth and fifteenth centuries was a particularly prolonged period of fighting when the French throne was claimed by Edward III. Although both sides made continuous raids on each other, the bastion of chalk Downs overlooking the channel was hardly affected. Even in the more recent times of World War I and II these Downs were only slightly marred in places by the construction of underground operations rooms, occasional radar and communications masts and a few white concrete roads for the military. The people of Sussex and Hampshire, mixed as they now were with Celtic, Roman, Anglo-Saxon and Norman blood, settled down into the feudal life of mediaeval England. The beautiful hills, wooded in the west and grass-covered in the east, dominated the little villages at the base of their northern slopes and overlooked the coastal plains in the south-west. To the east the great white cliffs looked across a more friendly channel protected by an English fleet that grew ever stronger under Henry VIII and Elizabeth I, eventually to dominate the oceans of the world.

The people of feudal England, then more than now, were occupied in farming; rearing sheep on the hills, and growing cereals and vegetables on the fertile lowlands. Watermills, introduced by the Romans, of which there came to be 157 in Sussex, ground the plentiful supply of corn. During the twelfth century many of these were replaced by primitive windmills that became a familiar sight as they turned briskly in the southerly breezes.

It was not until Tudor times that the mediaeval society slowly crumbled and the foundations of a true nation state were laid. Prosperity grew, stability and a settled way of life became the norm for the Sussex farmer. The Spanish Armada was defeated in 1588 and the English fleet continued to sail unchallenged. The Civil War of the seventeenth century interrupted this tranquillity but briefly. Ordinary Sussex folk who supported Parliament against a weak Charles I soon settled under a more democratic constitution and continued their daily lives much as usual. Along the coast at this time and for long after, Sussex men earned an unsavoury reputation for their heartless treatment of ships that foundered on southern shores. Once a ship was firmly stuck fast on rocks or sand, it was plundered unmercifully, especially if it was a foreigner. William Congreve in *The Morning Bride* aptly describes this fact.

> As Sussex men that dwell upon the shore
> Look out when storms arise and billows roar;
> Devoutly praying with uplifted hands
> That some well-laden ship may strike the sands;
> To whose rich cargo they may make pretence
> And fatten on the spoils of Providence,
> So critics throng to see a new play split
> And thrive and prosper on the wrecks of wit.

Throughout our history economic requirements have led to frequent changes in farming, from arable to pasture and vice versa. The landscape and character of the South Downs together with the make-up and variety of its natural history changed at the same time.

At the end of the eighteenth century, demand for grain, created by the Napoleonic Wars, resulted in large areas of downland being ploughed up for the first time since the Celtic age. The repeal of the Corn Laws in 1846 and the resultant sharp drop in cereal prices encouraged many farmers to revert back from arable to grass so recreating the sheep-grazed downlands. Imports of cheap mutton and wool from New Zealand towards the end of the 1880s made sheep farming less profitable, and by the time of the First World War sheep numbers on the Downs were down to about a third of their original level. The reduction in grazing allowed the encroachment of scrub, mainly *gorse, elder* and *bramble* on ancient grasslands inevitably leading to a reduction of the rich chalk flora, yet, at the same time resulting in an increase of birds like the *yellowhammer, linnet, whitethroat* and many kinds of *warbler.*

The need for cereals in the First and Second World Wars resulted in more downland being ploughed, and in 1945 government grants providing subsidies for cereal farming had a disastrous effect on many remaining areas of old chalk grasslands. Cheap artificial fertilisers meant that cereal-growing even on the poorest land became more profitable than rearing sheep. So the natural grasslands of our Downs began to disappear except on the steeper slopes which were impractical to plough. Scrub invasion on these remaining slopes continued during the post-war period due to reduced grazing, and the spread of scrub was given a boost in 1953 with the arrival of myxomatosis which decimated the rabbit population. In the 1960s growing public awareness of the need to preserve what was left of our natural heritage resulted in the establishment of conservation bodies such as the Sussex Wildlife Trust and English Nature. Sadly, as we shall see in subsequent chapters, some of the good work of these bodies is being eroded in the overgrazing of sensitive areas and by the reappearance of the prolific and voracious rabbit.

The destruction of the wildwood continued after the Norman invasion as more cultivated land was needed to support an increasing population. On the Downs the most heavily wooded parts are in East Hampshire and West Sussex where one can still find solitude and seclusion in the great forests of Westdean, Charlton and Houghton. Some notable woodlands can be found on the eastern downland such as at Friston but more often one will come across a lone wind-sculptured *hawthorn*, its branches growing almost horizontal away from the prevailing south-westerly winds. This 'slanting' of the branches is not because they are bent into this position by the force of the wind, but is a result of the young shoots only being able to grow on the sheltered side of the bush unhindered by the salt-laden and destructive gales. The large acreage of woods on the western Downs compared with the east is partly due to its wetter climate. Rainfall decreases from west to east as the rain-fronts exhaust themselves on their journey from the Atlantic, the Downs above Chichester receiving about 100cm a year while those in the east about 60–80cm. Another significant factor was of course the more intensive wood clearance on the eastern Downs for sheep grazing.

Until about two centuries ago many of the ancient wildwoods were managed under a coppice system, where the underwood was cut on a 7 to 20 year cycle depending on the species. Today coppice plantations are largely confined to the Weald where the *sweet chestnut* is grown for traditional Sussex cleft-rail fencing. In the 1950s and 60s, vast areas of coppiced woodland were grubbed up to make way for coniferous plantations of *Scots pine*, *Corsican pine* and *Japanese larch* which are used as woodpulp. Whereas the gradual clearance of woodland from the Downs over hundreds of years has given plants and animals room to spread and colonise the open spaces, the advent of chestnut coppices and coniferous forests has done little for woodland plants which find the suffocating leaf fall of the former and the dark interior of the latter unpalatable. On the other hand the wide grassy forest rides in these habitats do give rise to flowers such as the *devils-bit scabious*, *heath bedstraw* and *lousewort* and the coniferous trees attract specialist birds like the *crossbill, coal tit, goldcrest* and *robin*.

Water-meadows in the river valleys that cut through the Downs were once very extensive and harboured an abundant variety of indigenous plant, bird and animal species. However, these lush meadows have steadily been drained in the latter half of the last century to meet the needs of the farmer. So the once common birds such as the *lapwing, redshank, snipe* and *yellow wagtail* are now largely confined to the protected areas of Amberley Wild Brooks and the Pulborough Brooks Nature Reserve. Some plants too, like the *flowering rush, arrowhead* and *marsh marigold* have become scarcer as a result of the loss of their water-meadow habitat.

One of the greatest tragedies of the South Downs happened in the early part of the twentieth century when the landscape of the hills between Rottingdean and Newhaven was changed forever by housing development leading to the unsightly new towns of Saltdean and Peacehaven. Luckily the need to protect our natural downland from such development was soon recognised at this time and thanks to the bodies like the National Trust, the Sussex Downsmen and Eastbourne Corporation, such beautiful areas as the Seven Sisters and Beachy Head were saved and their beauty preserved for future generations.

I am very conscious that this brief history of our chalk Downs has not by any means covered every aspect of man's relationship with the environment over the past twelve thousand years, but I hope enough has been written to stimulate the reader in the quest for further knowledge. This second chapter has however shown how man has shaped and moulded the landscape into what we see today. Plant and animal life, although much depleted since former days, still give us a varied and interesting heritage. This heritage must be protected at all costs from further exploitation and destruction. As the Sussex Wildlife Trust in its book *Vision for the Wildlife of Sussex* concludes:

> There is a real choice to be made about how much wildlife there will be in Sussex in 2045. A choice about what Sussex will look like. A choice between a green and pleasant county, free from pollution and with growing wildlife numbers, or a declining countryside where the environment and wildlife are under constant pressure. The Sussex Wildlife Trust believe that there is only one possible choice.
>
> Dramatic improvements are possible when people and politicians put their efforts to solving problems. Those who can remember the dreadful urban smogs in winter in many big cities in the quite recent past know it is possible to solve environmental problems. Taking the action proposed in this modest and practical 'Vision' will produce a much enriched wildlife and countryside and add to everyone's quality of life. It will result in a fine Sussex which we should be proud to leave to our heirs.

With this vision of the future, come with me now as we look in detail at the natural history of the Downs in the following chapters.

Belle Toute in 1946

21

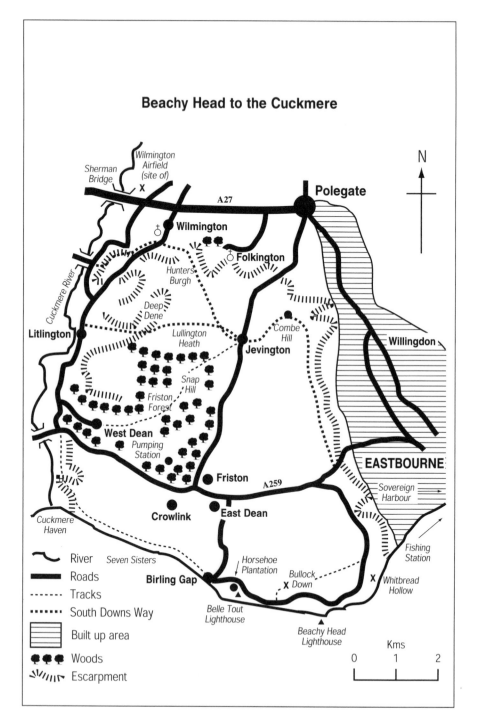

Beachy Head to the Cuckmere

Sherman Bridge

Wilmington Airfield (site of)

X

A27

Polegate

N

Wilmington

Folkington

Hunters Burgh

Deep Dene

Cuckmere River

Litlington

Lullington Heath

Combe Hill

Jevington

Willingdon

Snap Hill

Friston Forest

West Dean

Pumping Station

Friston

A259

EASTBOURNE

Sovereign Harbour

Crowlink

East Dean

Cuckmere Haven

Fishing Station

Seven Sisters

Horsehoe Plantation

Bullock
X Down

X Whitbread Hollow

Birling Gap

Belle Tout Lighthouse

Beachy Head Lighthouse

～	River	
▬	Roads	
- - -	Tracks	
•••••	South Downs Way	
▤	Built up area	
🌲🌲🌲	Woods	
ﾊﾊﾊ	Escarpment	

Kms

0   1   2

3

BEACHY HEAD TO THE CUCKMERE

Belle Tout

Let the long flat concrete promenade on Eastbourne's western sea front be our starting point; hardly inspiring to a lover of nature and the solitude of hills, woods and open fallow fields, but ignore the passing people, their chattering children and dogs straining at their leads and view ahead the distant cliffs and wooded downland slopes. Look also close by at the steep embankment where man's profligate planting of shrubs and flowers stretch above you in a vertical palate of colours that blend in so well with the blue sky and greyish greens of a billowing sea. Here amongst them you will find indigenous wild plants such as *bird's-foot trefoil* and *valerian*, which I well remember growing in the same place some fifty years ago when, as a young boy, I played on the pebble shore. Along the top path beside the main road a whole line of evergreen bushes grow about 2–3 metres high with slender purple green branches and tiny scale-like leaves. This is *tamarisk*, a native of south-west Europe, particularly Spain, its name deriving from the river Tamaris of that country, along whose banks the shrub still flourishes. It is not a native of Britain but is common in the south, south-east and west of England where it is used as a shelter hedge because of its power to withstand cutting salt winds. Its pale pink or white flowers are tiny and are borne in clusters on blunt spikes and appear from July to as late as November. The pliant stems were once used for making lobster pots, particularly in Cornwall and the Isles of Scilly. I remember it being quite common around the Eastbourne fishing station near Langney Point. I wonder if in the distant past it was used there for that same purpose?

At the western end of the promenade make your way up the steps to the main coast road and just past the last house on the left, take the coastal track that leads off towards the cliffs. A hundred metres on your right is a clump of bushes and trees; look on the bank in the shade under these bushes for the shiny green arrow-shaped leaves of the *Italian Lords-and-ladies plant*. This is a rare species related to the very common *Lords-and-ladies* flower, and is found in just a few places in Sussex on stony ground on hedge banks and near woods. It is a member of the arum family and is easily found in February when the undergrowth is absent and the leaves, covered with thick white veins, begin to appear. The leaves are followed in May and June by the flower stalk from which unfurls a bract that resembles a monk's cowl. Inside is a yellow cylindrical column called a spadix which puts out an odour of decomposing manure and attracts small flies called *owl-midges* (of the genus *Psychoda*). These flies then descend to the base of the spadix where the small male and female flowers are arranged around the central column. On their way the flies pass over a ring of stiff bristles which prevent them from escaping. If the flies have already visited a similar plant they bring with them pollen which then fertilises the female organ. The male flowers then discharge their pollen over the insects, the bristles shrivel up, and the flies are then free to carry their pollen to another flower. The spadix of the Italian Lords-and-ladies being yellow distinguishes it from the common species which is usually purple in colour, rarely yellow, and which flowers slightly earlier in April and May. The bright red berries of both species appear in July and August and are very poisonous and can be fatal to children.

In many places along the track are large patches of a tall plant with glossy green leaves, which is often the first new foliage of the year to appear on hedge-banks near the sea. This is *Alexanders*, whose stout, rounded, solid stems bear large umbels of small yellow flowers from April to June. This plant was probably first introduced to this country by the Romans from its native Mediterranean habitat. It was used to add myrrh-like flavour to broths and stews and to be eaten raw in salads. Later it was planted in early monastery gardens and became a favoured plant of cottage gardens. The stems, if cut down and trimmed into pan-sized lengths, make a delicious vegetable. Peel the stems with a knife and then boil them in salted water for about 8 minutes until tender. They are then best eaten with black pepper and butter. Alexanders is very common here on the south coast and has even established itself on roadsides in the centre of towns such as Eastbourne. The Latin name for Alexanders, *Smyrnium*, is derived from the Greek word smyrrna, in allusion to its odour.

Further along the path you will see on your right a patch of exposed chalk and here in June and July you will see the tall plants of *red valerian*. Nearly one metre tall and hairless, it has red or white flowers borne in clusters at the end of the stem and lance-shaped grey-green leaves. Introduced as a garden plant from the Mediterranean in the 16th Century, the red valerian has now spread to most of Britain and Ireland. The leaves are bitter but, if cut young, can be used in salad. The whole plant has a faint fragrance and is much visited by butterflies and other insects.

Suddenly, as you reach the brow of the hill a panorama of colours appear, the blue sky, deep green sea and downland grass bespattered with large expanses of bright yellow. This is Whitbread Hollow and the colour is caused by the *bird's-foot trefoil* plant which carpets our meadows and roadside banks with masses of deep, golden yellow pea-like flowers in early summer. Red streaks often appear on the back of the petals and guide bees to the nectar deep in the throat of the flower.

Wherever you find the bird's-foot trefoil you are likely to come across the *common blue* butterfly whose small green caterpillar feeds on the leaves. As its name implies, the common blue can be found throughout the British Isles and is the most prevalent of the twelve species of blue butterfly. The male common blue is the more striking of the sexes with its metallic blue wings that glitter in the summer sun; both sexes have white margins around the edges of their wings but the female can be distinguished by her brown colour and by the addition of orange markings. The female also occurs in a blue form similar to the male but still has the distinguishing orange markings around the wings, especially on the hind-wing. Look carefully down there on that trefoil plant, there's a female common blue just laying her egg. The tiny eggs are laid singly on the upper side of the leaves and hatch out into small green caterpillars after about a week. Caterpillars feed on the host plant and take about 45 days to become fully grown, before turning into a chrysalis. Many caterpillars spend a longer time feeding and hibernate during the winter. The chrysalids, when formed, last about two weeks before the adults emerge in late

April, May or June, two generations usually being produced each year. Adult blue butterflies, like many of the smaller species, live for only about two to three weeks, so this female having laid her eggs and perpetuated her species will not live for many more days. Look carefully as she spreads her wings to reveal her beauty before flying off.

When the sun is shining the common blue is a difficult butterfly to get near to, but when it is cloudy it is often possible to pick them up out of the grass; they like the warmth of one's hand and can be easily seduced to perch on one's finger. As the sun goes down you may see groups of butterflies clinging to grass stems resting with their heads pointed downwards; they are roosting for the night.

Heat is very important to butterflies and in order to fly they need the sun's warmth to circulate the blood in their veins. Moths, too, require heat to fly and since most fly at night they can sometimes be seen vibrating their wings furiously, so increasing their body temperature by muscular activity. The wing scales of moths are thicker than butterflies as a rule and they act as an insulating 'fur' to retain body heat. It is fairly easy to distinguish butterflies from moths. Most butterflies fly by day and have distinctive colouring on their wings, whereas moths generally fly at night and are dull in appearance. Most butterflies sit with their wings together over their back whereas moths rest with their wings spread out. There are many exceptions to this general guide and the only true method of distinguishing between them is to examine their antennae. These antennae are always club-tipped in the butterfly, while those of the moth are feathery and finely pointed.

Growing amongst the grass by the cliff edge is a tall stately plant with bell-shaped blue and purplish-red flowers. Occasionally you may find a plant with white flowers, a magnificent sight. This is the *viper's-bugloss* which can grow up to a metre tall and is common on top of the Downs near the sea. It has rough, hairy stems and long, narrow leaves and can be found flowering from May to September. The flowers are full of nectar and attract many varieties of bees, butterflies and moths. After pollination each flower produces four nutlets resembling a viper's head and for this reason it was said to be a cure for snake bite. It was also used as a cure for melancholy.

Two rare plants associated with Whitbread Hollow are the *early gentian* and the *sea radish*. Only recently discovered in this area, the early gentian can be found flowering in the middle of June to the end of July amongst the short grass on both sides of the

26

scooped-out valley of the Hollow. Some years it is absent but in others there are literally thousands flowering amongst the bird's-foot trefoil and other downland plants. *Early gentian* is a small jewel of a plant, only 5–15 cm tall. The pale purple flowers with four or five sharply pointed petals are borne at the end of flower stalks, the main branches of which grow out from the base of the plant. The basal leaves of early gentian are spoon-shaped, whereas those growing on the stem are lance-shaped and grow opposite each other. This gentian is one of six other members of the same genus and as we shall see in Chapter 5 there is another similar species, *autumn gentian*, which is relatively common and flowers later in the season. The gentian family honours Gentius, a pirate King of Hungary, who lived in the second century BC and is believed to have discovered the medicinal uses of plants. The roots of gentian were used in mediaeval times to produce a bitter substance which when added to other medicines acted as a tonic.

Growing out of the chalk cliff itself and in the cliff-top grasses is the handsome *sea radish*. Up to a metre tall, this hairy plant has four-petalled yellow flowers which bloom from June to August. It can be distinguished from the *wild radish* by having only up to three joints in its pod-fruit as opposed to up to eight. The wild radish with its paler flowers does not like chalky soils but can be found as a persistent and common weed in arable fields throughout the British Isles. The seeds of the sea radish are dispersed by the sea on which they will float for several days, and interestingly this rare Sussex plant can also be found growing some miles to the east of Whitbread Hollow at Normans' Bay, the seeds probably carried there by the prevailing sea currents.

Our journey now takes us up and over those great cliffs of Beachy Head and down towards the old lighthouse of Belle Tout. I remember walking this route just after I had started writing this chapter. It was early on the morning of the 11th January 1999. After a prolonged period of rain, the day had dawned bright and clear with just the hint of a light onshore breeze blowing from off the sea; a light overnight frost had coated the downland grass with a covering like icing sugar. I heard the birds first, the cries of startled gulls and the harsh cackle, 'tchack, tachack' of disturbed *jackdaws*. Something was clearly amiss. I topped a small rise and then could see, just beyond and to the west of Beachy Head light-house, a huge fall of chalk cliff stretching across the beach to the base of the lighthouse shelf. In front of me, a 100-metre stretch of the cliff top had sheared away completely to the sea below. Clearly it had happened not many hours before and I could imagine the scene; first a slight rumble, then a gradual and sustained roar as the huge bulk of white chalk and flints which for thousands of years seemed so secure high above the waves, fell earthwards to be broken into pieces of jagged rock. Incredibly, these would in time be worn down into rounded boulders by the repeated action of the tides and pounding water. As I watched the slow incoming tide wash over and erode the extremities of the chalk fall, I wondered at the marvel of it all; here another piece of England broken away to be dissolved into a nothingness. A few shallow cumulus clouds formed, their colour darkened by the low sun in the south-east. The jackdaws, their plumage matching almost exactly the colour of the clouds, cackled incessantly, thus completing a melancholy backdrop to this sad scene. 'A good job I wasn't on the cliffs a few hours earlier', I mused as I contemplated what I knew must be the largest cliff fall in living memory.

Belle Tout lighthouse appears ahead as a modern dwelling place reconstructed on several occasions around and on top of the original light built in 1828. As a lighthouse it was never very successful because it was often obscured by mist and in between 1899 and 1903 at the insistence of Trinity House, a new 35-metre tall Beachy Head lighthouse was built at the base of the cliffs. The area of downland immediately around Belle Tout was once an Iron Age camp and in 1968 was identified as the site of ancient Beaker people, the only one of its kind in south-east England. In February 1971 a cliff fall just below the lighthouse revealed a 60-metre deep well-shaft running vertically down to beach level; it had footholds placed neatly into its sides and confirmed a long-term habitation of the site by early man.

The area around Belle Tout provides a suitable habitat for four interesting plants, the *field fleawort, early purple orchid, early spider orchid* and *small hare's-ear*.

The *field fleawort* as we mentioned in Chapter 1 has a peculiar propensity to grow near Iron Age settlements. The reason for this is not clear but perhaps it was used to ward off fleas or for a medicinal purpose – the name 'wort' in plants usually signifies this. Field fleawort is easily recognised by its rosette of broad, oval leaves covered in woolly hairs, from the centre of which arises a single, also woolly stem. The top of the stem, some 10–30 cm high, terminates in a cluster of 1–6 yellow flower heads branching from one point. Look for this plant where the grass is reasonably short, and in May when the flowers are at their best.

The *early purple orchid* is an ancient woodland indicator plant and one might wonder what it is doing here, flowering on top of this dry, chalky hilltop just to the west of the lighthouse. Probably it flourished on this spot a thousand years or more ago when it would have been heavily wooded with *ash*, *hazel* and *pine*, and over the years as the wood and scrub was cleared it adapted to the changing habitat. The early purple orchid can be identified by its rosette of glossy blunt leaves with round black spots and blotches. The stem, up to 60 centimetres tall, ends in a loose spike of six to thirty red-purple flowers which bloom in April and May. When they first open, the flowers have a delicate fragrance but this soon changes to an 'odour of cats' or goats. This orchid was mentioned by Shakespeare in *Hamlet*, where he refers to it as 'long purples' and 'dead men's fingers', one of the earliest references to the plant. During the nineteenth century poorer people ground up the dried tubers of this orchid, mixed it with hot milk, honey and spices and drank it as a breakfast coffee; it was called 'salep' and was thought to have aphrodisiac properties!

On the path just below Belle Tout and to the east of Horseshoe Planation, look out in April and May for a diminutive plant of the orchid family, the *early spider orchid*. I remember this place well when as a boy in 1952 I had searched all day for this orchid amongst the rabbit-grazed downland above Eastbourne. I knew the habitat was suitable as it was mentioned in A. H. Wolley-Dod's *Flora of Sussex* as growing around 'the Beachy Head range in several localities'. It was one of the last days in April, the sun shone brightly out of a cloudless blue sky, a pair of *peregrine falcons* soared high above the white cliffs, their calls of 'zee zee zee' carried far by a gentle southerly breeze. An *adder* with its zigzag markings, recently awakened from its hibernation, slithered through the *cowslips* that covered parts of the slope with patches of yellow. I remember cycling home after a fruitless search and then on a whim I decided to look along the path below the lighthouse. I placed my bicycle on the ground and almost immediately saw for the first time the object of my search. The beautiful 6-centimetre-high orchid was exquisite, with a broad, brown labellum just like the body of a bloated spider, with two globules of nectar shining out from the flower like a pair of eyes. I remember my excitement as I took picture after picture with my battered old Brownie camera.

The early spider orchid is perennial and tends to grow in colonies not far from the sea. In some years there are no flowers to be found, and a year or so may pass before they reappear in all their glory. The reason for this is unknown but as we shall see later with the *autumn lady's tresses* orchid, it may have something to do with the amount of rainfall over winter and early spring. Like many of the orchid species the early spider has a root system which consists of two egg-shaped tubers which act as a food source. An interesting fact about orchids is that in many cases the seed must come into contact with a root fungus called mycorrhiza before it can germinate. Although mycorrhiza is not uncommon, existing in roots of trees and plants, the dependence of orchids on this fungus probably explains why so many of them are uncommon. The three or four leaves of this orchid form as a rosette in late winter. They can be found, often with the edges burnt brown by frost, in February and March, their well-marked veins and grey-green colour standing out clearly from other vegetation in the short grass.

The early spider orchid is quite common on the chalk downs of France but sadly, like so many of our orchids in Britain, this species is becoming rarer, possibly because it is on the edge of its range here in England and even small changes in rainfall and temperature could affect its distribution.

The last of the four plants, the *small hare's-ear*, can be found within a metre or so of the cliff edge just to the east of the car park below Belle Tout. You will need patience and a good head for heights to find this tiny plant, seldom more than a few centimetres tall and flowering in June and July. It has sharp-pointed spoon-shaped leaves and the minute yellow flowers, borne in groups at the end of short stems, are enclosed by egg-shaped bracts. This annual plant is very rare and apart from this site is found in Britain only at Berry Head in Devon. If you find it, therefore, your pleasure will be complete.

Before walking across to the Seven Sisters a kilometre or so to the west, look across the tarmac road to Bullock Down which is a special site for butterflies, birds and wild flowers. This is a place still frequented by the *skylark* and *meadow pipit*, where the downland slope is just as it was in the past, unchanged by man's interference except for carefully controlled grazing by cattle and sheep. Here, amongst the lush grasses from April onwards can be found plants such as the *cowslip, early purple orchid, early spider orchid, greater knapweed*, the rare *moon carrot* and *dyer's greenweed*, to name but a few. In an age when money is all-important and when developers soon ravage and devastate our ancient, most pleasing and outstanding wildlife sites, this is surely a place that must be protected forever. Inevitably, of course, it will succumb to the ever encroaching sea. One only has to look down at Eastbourne from the hills above to see for oneself what man has done to that great shingle bank, The Crumbles. A place I knew as a boy where *little terns, common terns, wheatears, lapwings* and *ringed plovers* all nested and where many uncommon plants, such as the *man orchid*, and *soapwort* could be found, The Crumbles was then the second largest shingle bank in the world, Dungeness in Kent being the largest of all. It has now been developed, undoubtedly for the pleasure of many, into a marina called the Sovereign Harbour. The few acres of original shingle still harbour rare plants and as I write one can still find *bee orchids* almost 60 centimetres tall; but for how long will this unique environment survive? Already there are plans to enlarge the marina and build some 200 houses on the site!

The Seven Sisters, that long expanse of white cliffs cut so neatly by the sea into enormous sections of seven waves of hills, is a place of great interest to the naturalist. Our walk takes us over wide grassways starting at the first 'Sister', Went Hill Brow and in turn passing over Bailey's Hill, Flagstaff Point, Brass Point, Rough Brow, and Short Brow until finally ending up on the top of Haven Brow, the highest point of the group with its magnificent view over the Cuckmere estuary. How often have I sat here amongst the downland flowers and watched early summer sea mists creep up and dance above these scalloped cliffs and seemingly with their soft embrace touch the very heart of nature itself.

Often in January and February when low dark clouds of rain blow in from a heavy and menacing sea, a sense of melancholy descends upon the scene. The shore itself looks the colour of dark umber, draped as it is with fronds of seaweeds like *bladderwrack* and *serrated wrack*. Suddenly you see a group of birds wheeling and diving in the wind that sweeps over the white cliffs; *fulmar petrels*, distinguished from gulls by their heavier rounded heads, tubular nostrils and short thick necks. The body, head and fan-shaped tail of the fulmar is the colour of ivory which contrasts with the light grey of its back and upper wing surfaces. These birds are awkward when on land and quite unable to stand, but when airborne they display a masterly skill of gliding low over the water or along the cliff faces. Here on the Seven Sisters they seem to hover and dance on the wind as it sweeps over the cliff edge and seem curiously interested in humans, perhaps because they spend most of the year in the middle of the Atlantic feeding on plankton and small crustaceans or fish brought up in the ocean currents. They come back to the coasts in January and February to find a mate and to lay a single white egg in April or May in a simple unlined scrape on any convenient ledge or cavity on the cliff face.

They nest in small colonies and incubation is shared by both sexes over a period of about 50 days. The chick obtains its food of regurgitated fish by placing its bill inside that of the adult. The young gain weight rapidly and after about 45 days they are abandoned by the parents. A few days later they fly down to the sea, and eventually they make their way out to the vastness of the deep ocean; whether they ever join up again with their parents is unknown.

The adult fulmar, when incubating, has a habit of ejecting foul smelling oily green liquid at any intruder that disturbs it and this is a very effective defence mechanism against other bird predators such as the black-backed gulls and even rats. The name fulmar or 'foul gull' is possibly derived from this habit. The fulmar's spread around the coasts of Britain is a great success story as in 1878 it was only known to breed in Shetland. It came to the cliffs of Sussex in about 1965 but breeding here was not proved until 1976. The population has since increased every year until now there are about 150 pairs nesting along the chalk cliffs of the county. The spread of the species has been put down to change in fishing practices where fish are now gutted at sea and the offal is thrown overboard as a plentiful and easily obtained source of food for the bird. I hope the fulmar continues to thrive here on the Seven Sisters and to give pleasure to those who have the time to watch and enjoy its antics.

From on top Flagstaff Point when the tide is low, look down to the sand just to the west of the farthest rocks and there you will see the aged wooden ribs of the *Coonatta* sticking grotesquely out of the beach. The shape of the hull of this sailing barque can be clearly seen, a sad fate for a vessel that had set sail from Sydney, Australia, with a cargo of copper and wool in 1876. The remains of other vessels can be seen under the cliffs, notably the SS *International* (1899) and the German submarine of 1917, but little is left of such ships as the *Nympha Americana* (1748) or the *Golden Fleece* (1778).

As you walk over 'The Sisters' you will not fail to notice the large numbers of sheep that graze on the wide expanse of grass. Placed all along the edge of the cliff about 30 metres inland is a metre tall wire fence to prevent the sheep from accidentally falling over the edge. Interestingly, this has created an area where plants can grow undisturbed. Over the past 10 years sheep, and at times cattle, so heavily grazed the top of the Seven Sisters that plants that were once common here, such as the *viper's-bugloss* and *cowslip*, have all but gone. However, inside the new fencing the

32

grass is thicker and one can find them and others like the *field fleawort* growing in abundance. It is a sad reflection on the management of such grasslands; for what was once a haven for the *meadow pipit* and *skylark* and rare plants, has now become a barren wilderness of close-cropped turf. Good management is called for, to control grazing animals in numbers and by season. (Recently I counted over 300 sheep on just one hillside here). Why cannot they graze on botanically rich areas such as this between October and the following March, when most flowers are absent and birds are not nesting? Later in this book I will describe other sensitive areas where overgrazing exists with the objective of bringing the problem to the attention of those responsible.

If you walk over the Seven Sisters in the spring or autumn, look out for two of our migrant butterflies, the *red admiral* and the *clouded yellow*. For many years the idea that butterflies migrated was hotly debated but in the middle of the last century there was adequate proof that some species were able to fly long distances and at surprising average speeds of some 15 kilometres per hour helped by favourable winds. The eggs of some migrant females do not develop until after their long journey when their energy supplies have been replenished. Little is known of their actual movements across the ground and sea, how much rest they need and their 'decision-making', but clearly much must be dependent on the temperature and wind conditions at the time. For most butterflies, migration is a one way journey; unlike birds they do not attempt to return. However there are exceptions to this rule as shown by two butterflies mentioned earlier, the red admiral and the clouded yellow.

The *red admiral* is one of our most beautiful butterflies with its distinctive red bands and white markings on a black background. The first migrants arrive from across the Channel in May and the female lays her eggs singly on nettle plants. These hatch after about a week and if you look carefully at a patch of nettles you may see the brown or green caterpillars (there are two types) feeding on the leaves from under a protective 'tent' which they have made by drawing the edges of two leaves together with silken threads. After about 30 to 40 days each turns into an attractive gold-spotted chrysalis from which the butterfly itself emerges after another 16 to 18 days. There are often two broods a year and in the warmer southern parts of Britain adults can survive for 8 months and more. Although a fast flier this butterfly is easily seen except when it is resting with wings folded amongst vegetation, when its marvellously decorated undersides blend in well with its surroundings. It is also territorial and can often be seen chasing away other intruding butterflies from its patch of hedgerow, woodland or garden. In the autumn large numbers congregate and feed off windfall fruit rotting on the ground. Unlike other butterflies, such as the *small tortoiseshell* and *peacock*, the red admiral has not learnt to hibernate and once the colder nights draw in, most die. However, I have often seen small groups of somewhat ragged-looking specimens flying southwards over the Seven Sisters in October and November, obviously trying to escape the onset of a British winter.

Another migrant butterfly that is much less common than the red admiral is the *clouded yellow*. This beautiful orange-yellow butterfly with black borders to the upper sides of its wings comes to us from North Africa and Southern Europe where it breeds continuously. Every spring it migrates northwards and when the winds and climatic conditions are favourable it will arrive in Britain at the end of May. Often the migration loses momentum in central France but exceptionally they arrive in huge numbers, as they did in 1947, 1955, 1983, 1992 and 1994. What a thrill it was then to see them fast flying in flashes of chrome yellow across the South Downs. The butterfly feeds on the nectar of common plants such as clover, thistle and knapweed and is a prolific breeder in warmer countries producing up to four broods a year, but here in Britain only a single brood is produced. None survive the cold damp weather beyond November and they can sometimes be seen in large numbers flying south in the autumn trying to cross the channel to the warmer latitudes.

Our journey now takes us away from the cliffs and inland, through one of the finest British beechwoods in the country, Friston Forest, and then on through Jevington to Combe Hill which overlooks Polegate and Willingdon.

As you make your way to the southern edge of the forest past Exceat New Barn consider how these sloping hills and gentle valleys looked a hundred years ago. Since the late Bronze Age (600–500 BC) they were tilled by simple ploughs drawn by two oxen; gradually a system of ploughing evolved where a team of six or eight oxen pulled a two-wheeled wooden 'Sussex Plough'. This plough, refined as it was by Sussex farmers over the centuries, proved efficient on undulating ground with steep-sided slopes and thickly strewn with flints. That great Sussex writer on nature, Richard Jefferies (1848–87) described them thus: 'There was hardly a more imposing spectacle than of these quiet beasts of burden stepping slowly forward before the plough against a background of grey sea and sky, their huge horns sweeping forward in generous curves.'

Until about the middle of the nineteenth century all the draught work on downland farms was done by oxen, either the native red Sussex breed, or the Welsh blacks. Sadly (for us) the sight of these patient creatures, slow in movement, but unceasing, peaceful, and one might say, melancholy in appearance, is no longer with us, the last working team at Exceat having ceased in the 1930s.

Friston Forest beckons us with its embrace of tranquillity and peace, a place where one can forget the troubles and anxieties of all this unintelligible world. Here amongst the tall, straight beech trees one can feel the hand of nature, for nature is all about. One can touch and see too, the timeless yet changing mood of the forest; from the silence of deep snow in winter and the purple tinge of its trees before they leaf in spring, to its shady coolness in the summer heat and the kaleidoscope of autumn colours which beckon the onset of winter, completing the wondrous cycle of nature.

Friston Forest dominates the area between East Dean and Jevington with its 1,500 acres containing over four million trees, mainly beech. Originally planted in the 1920s and 1930s by the Forestry Commission with the intention of creating a beech forest, it rapidly grew into a coniferous woodland which was grown as protection for the slower growing beeches. After much criticism, the conifers were gradually cut down and sold for pulp and now this is one of the largest beech woodlands in the country.

As you descend the track from the main road look out in February and March for the first of the *early wood violets* to be followed a month or so later by the *common dog violet*. The two plants can be distinguished from each other by the colour and nature of the flower spur; in the case of the early wood violet the spur is a darker lilac colour than the petals, and in the dog violet the spur is broader and furrowed and is paler than the petals. Another plant of the same family flowering in March and April is the *hairy violet*. This species, as its name implies, is covered in hairs and this, together with the fact that all its leaves and flower stems arise from the base of the plant, distinguishes it from the other two. All three can be found growing in the same locality on the grassy banks of the chalky tracks through the Forest, but the hairy violet is by far the commonest violet to be found on the chalk, particularly on open ground.

When spring gives way to early summer and the beech leaves are fully developed, the interior of the forest takes on a dark and somewhat uninteresting appearance. But do not be deceived; let your eyes wander to the dimly lit spaces under the trees, for here two orchids can be found, namely the *white helleborine*, and the *broad-leaved helleborine*. The former will be described more fully later but briefly it is 15–60 centimetres tall with fresh green egg-shaped leaves and creamy white flowers with a yellow centre which bloom in May. Later in July and August the tall spikes of the broad-leaved helleborine can be found in the more mature part of the beech forest. Up to 50 centimetres tall, it has an imposing flower spike consisting of up to a hundred pink and brown flowers, pollinated almost entirely by wasps, which often appear in large numbers covering the spike all over, in late August. They do their work efficiently as can be seen by the number of pollinia that have been removed from the flowers, but like other helleborines the seeds take at least eight years to develop into flowering plants.

Friston Forest is the place to find the *yellow bird's-nest* plant (not related to the orchid with a similar name) for here is the largest concentration of the plant in the British Isles. This sapro-phytic plant relies on the benevolent mycorrhiza fungus for its nutrition; this fungus feeds the plant with nutrients by breaking down the constituents of rotted beech leaves. The yellow bird's-nest grows through this beech litter and puts up its pale yellow fleshy stem some 20 centimetres high in May and June. The drooping dingy yellow fragrant flowers appear in July and, after fertilisation, they stand out horizontally with only the uppermost ones becoming erect. Look out for this unusual plant in the darker areas of the forest just north of the track from the Pumping Station to West Dean.

Another uncommon flower to be found here is the *pheasant's eye*. It is one of the many colourful annuals that has been all but eliminated from our cultivated fields in the last fifty years by herbi-cides and modern farming methods. Once this attractive plant with its crimson red petals and dark purple centre was a common weed in the chalky cornfields of Sussex and Hampshire, but now it is seldom seen. In the eighteenth century it was gathered up by local people and sold in Covent Garden market as 'Red Morocco'. The striking flowers are set off by the background of its deeply cut and divided pale green leaves. Its Latin name, *Adonis*, is an ancient one and is said to be derived from Adonis, who in Greek mythology was killed in the forest by the wild boar; his blood was said to have fallen on this plant. It flowers from late May to August on disturbed ground in the forest particularly near the new car parks and on the chalk banks of some of the wider tracks. Interestingly every 14 or 15 years it appears in huge numbers on fields around Crowlink near the downland village of East Dean; the last time it did so was in June, 1996.

Butterflies abound on the sunlit tracks of the forest and on a hot day in July you will almost certainly see the *small tortoise-shell, speckled wood* and *meadow brown* to name but a few. But look out for the once scarce *white admiral*, now seemingly becoming more common, particularly in the northern part of the forest near Lullington Heath Nature Reserve. The white admiral has distinctive black wings with a white band running across them, and can more often be seen feeding on bramble flowers, their favourite source of nectar. The female butterfly seeks out the *honeysuckle* on which to lay her eggs in July, for this is the sole food plant for the caterpillar. Having fed on honeysuckle leaves until autumn the caterpillar hibernates throughout the winter within the crinkled leaf, emerging in the spring to feed on new fresh foliage. It then turns into a chrysalis from which after about 12 days the butterfly emerges in June. Like many butterflies the life span of the adult is quite short, in the case of the white admiral only about 25 days. The key to the success of this butter-fly is woodland management for it requires shade and an abun-dance of old honeysuckle with its long trailing strands. The great storms of 1987 and 1990 created havoc in many southern wood-lands and great gaps were torn in what had been closed woodland canopy, creating open woodland conditions. This opening of the canopy led to a decrease in the white admiral population, which has only now started to recover with the planting of new trees and the growth of others. Once you have seen this lovely butterfly soaring up into the trees through the leaf canopy and then swooping low to feed briefly on bramble flowers you will never forget it.

One species that has benefited from the opening up of the woodland is the *grizzled skipper* which likes open areas particularly with some patches of bare ground where it can bask in the sun's reflected warmth with wings outstretched. This conspicuous and enchanting little butterfly is easy to identify with its dark colour and white irregular spots on its upper surface and a black and white chequered fringe to the wings. The caterpillar's main food plants are *barren strawberry, creeping cinquefoil, wild raspberry* and *silverweed*. These are common plants of the open trackways through the forest so here this species has perhaps its greatest concentration in Sussex. Look for it in the early months of the year, April and May, as it flies with fast wing beats or rests in the sunshine with its attractive wings wide open.

Many of our garden birds are to be seen in the forest and one of the commonest must be the *blue tit*. The only member of the tit family with a bright cobalt blue crown to its head, this little bird is familiar to all who place nuts on a garden bird table, for it has now become a truly suburban species. However, it is equally at home in broadleaved woodlands where mature tree cover is greatest. In winter, blue tits form loose flocks when they search for insects and beech nuts among the dead leaves. Their cries of 'tsee-tsee-tsee-tsit' bring a cheerful note to the sombre quiet and bleak nature of the damp forest in midwinter. In the spring the birds pair off and in the forest they look for any likely hole in a tree, wall or even hedgerow bank. Its nest is made of grass, moss, hair and wool with a lining of feathers, and from six to as many as twelve white eggs lightly spotted with red are laid from April to June. It sits close on its nest and, if disturbed, will often hiss sharply at the intruder like a snake.

Another common member of the tit family, the *long-tailed tit*, can be seen flitting in small flocks through the bare undergrowth of the winter woodland. Their distinctive call, 'tzee-tzee-tzee' is sharper than the blue tit's, but the bird is easily recognised by its black and white and pinkish plumage and distinctive long graduated tail which is longer than the bird itself. Generally it feeds on insects but will take just about anything it can in a cold winter, searching for grubs and crysalids amongst leaves, cones, and seed pods of bushes and trees. The number of pairs available for breeding in the spring is greatly dependent on the severity of the winter. In the harsh winter of 1963 this bird was all but eliminated but the numbers have recovered well and it is estimated that there are now up to 30 pairs per square kilometre in our broad-leaved woodlands. The long-tailed tit is a master builder, skilfully constructing a domed nest in a thorn bush between a metre and five metres above ground, this being made of moss, wool and spider's webs lined with feathers, and its outside covered in lichens giving it a wonderful camouflage. Some seven to ten white eggs speckled with tiny reddish-brown spots are laid as early as March, with repeat clutches in May and June. Sometimes all the eggs hatch out and it is a marvel how the large brood of young can be contained in such a small and delicate nest; even more extraordinary is how the parents keep the chicks fed with insects during the three weeks or so they are in the nest. The large brood clearly helps this bird to survive as a species during a cold and prolonged winter.

When I wrote *A Natural History of the Cuckmere Valley* in 1996, I mentioned the *hobby* as a bird often seen in the forest that might soon be nesting there. There is now clear evidence that this bird is indeed breeding in the area; so watch out for this fast-flying bird of prey with its scimitar-shaped wings. You are more likely to see this beautiful falcon from mid-April to the end of May, just after it has arrived on its long migration flight from central and southern Africa; the first indication of its presence will probably be its call, a clear repeated 'kew, kew, kew' and a rapid 'kikikiki'. What a wonderfully evocative sound to hear, and then to see the bird itself, particularly on a summer evening as it swoops to catch its prey, typically a young swallow or dragonfly in flight.

Our walk through the forest takes us up north and slightly east to Snap Hill and the quieter parts. Here in May and June the environment of the wide grassy rides seems suffused with the silent stillness of the woods and on windless days an all-enveloping warmth dulls the senses and gives one restfulness and peace. What a place to stop awhile and reflect on all the problems of this unfathomable world, a place where one's very being seems attuned to nature and where the mind can be cleared to reflect on the meaning of life itself. Such places are so essential to man as a refuge from the stresses of a complicated and ever-busy lifestyle; we lose them at our peril, especially when so few exist in the increasingly suburban character of south-east England.

We reach the top of the brow where the 'gallops', a wide track used for training racehorses, overlooks the little village of Jevington. Across the valley lies our next destination; Combe Hill, standing proudly above the downland landscape beckoning one to visit and discover its history. But before taking the track down to the village look for an interesting small tree growing just on the edge of the forest beside the gallops, with greenish-grey, small oval leaves and small greenish flowers. This is the *spindle tree* which is quite common on chalk-rich soils in southern England. The tree seldom exceeds five metres in height and can be easily recognised by the fact that its younger twigs are dark green and square in cross-section whereas the older stems are rounded with pale grey-brown bark. The flowers appear in May and these, after pollination, turn into beautiful autumn fruits; bright pink four-lobed capsules which split open to reveal orange pulp surrounding a hard seed. Birds find the fruit irresistible and the seeds are widely spread around the countryside as a result. The spindle tree was so named because for centuries it was used by womenfolk (known as spinsters) for spindles to spin wool on account of the fact that its wood was hard, smooth and kind to the fingers. The wood is also used by gypsies to make clothes-pegs, skewers and knitting needles.

Take the gently sloping path through growing corn towards St Andrew's Church in Jevington. This church has a late Saxon flint tower crowned by a typical Sussex cap. To the north of the church lies Jevington Place in whose grounds grow some fine *lime* trees. These trees with their great height, pleasing outline, grace and elegance were commonly grown in the nineteenth century to enhance the beauty of country estates. They are easy to manage and their flowers are full of nectar which is attractive to honey bees; their flowers, when dried, make a pleasantly flavoured tea. Wood of the lime tree is creamy-white and firm and is easily sculptured into carvings. It is also used for hat blocks, shoe trees and piano keys. The branches of the lime trees here in Jevington are pitted with regular lines of holes which have been punched into the bark by woodpeckers seeking to suck out the sap.

The churchyard on this still spring morning is splashed with colour, small groups of white from dying *snowdrops* fade into a background of burgeoning green, and bunches of yellow appear from newly emerging *daffodils*; an exotic cherry tree throws out its blossom of pale pink flowers giving a touch of the orient to the scene. Look out for one of the rarest weeping elms and see its elegant branches curve so neatly to the ground. This tree, only about 6 metres high, is probably a hybrid of the *Camperdown elm* that originated as a seedling at the castle of that name in Angus, Scotland. The silence of the churchyard is suddenly broken by the rapid drumming of a *great spotted woodpecker* on a nearby dead branch. This drumming is a territorial sound giving a warning to other woodpeckers to keep out of the area. Looking in the direction of the sound with binoculars you see the male bird with its black crown and back, large white shoulder patches and red nape. This woodpecker almost became extinct early in the nineteenth century but has since recovered in numbers and has now become a common bird of our woodlands, extending its range throughout Britain. Of the other two woodpeckers you may see in the area, the *green woodpecker* is distinguished by its greenish back, crimson head and its laughing call; while the *lesser spotted woodpecker* is sparrow-sized, has barred black and white upper parts with the male having a crimson crown and the female a whitish crown. All three woodpeckers nest in holes dug out or enlarged in a tree and their four to nine pure white eggs are laid in April and May on wood-dust and chippings at the bottom of the hole some 25–50 centimetres deep. The eggs hatch out after about 16 days and the young are fed various types of insects (ants are favoured by the green woodpecker winkled out from their nests by the bird's long tongue). The nestlings fly from their nests about three weeks after hatching. The greater spotted woodpecker has a reputation for raiding the nests of other birds looking for young nestlings which they devour with relish; a common practice with many of our best-loved birds!

The sunken lane that leads down from the church has old flint walls covered with moss and three interesting ferns, *common polypody*, *maidenhair spleenwort* and *wall-rue*. What an attractive sight they make, especially when seen in the light of an early morning sun which shows up the delicate and intricate outline of their leaves against the uneven protrusion of ancient flint. Ferns, in fossil form at least, date back some 300 million years or more, when tree ferns of 30 metres high formed the world's finest forests. The wood of those early tree ferns are the basis of present-day coal deposits. Some 50 different species of ferns still grow in Britain but several have become rarer because of atmospheric pollution. Ferns are flowering plants that reproduce by means of minute spores found in sacs on the underside of the leaf; the spores germinating to form both male and female organs which fuse with the aid of moisture and develop into new young fern shoots. Ferns are very common in western Britain where the high humidity ensures the dampness necessary for fertilisation; in eastern England ferns are less prolific, but here on the shady north side of a village wall they seem to flourish.

The *common polypody* has distinctive fronds which are deeply cut into narrow, blunt lobes. These fronds are produced in early summer among older darker ones from the previous year. Here on the wall the plant is somewhat stunted and is only about 15 centimetres long, whereas when growing on rotten tree stumps in damp woods the fronds can be as long as 50 centimetres.

The *maidenhair spleenwort* is a pretty little fern with slender fronds some 5–20 centimetres long. The small oval dark green individual leaves that make up the fronds are fastened in pairs at equal intervals up the main stem which is wiry and black in colour.

The *wall-rue* is a small fern little more than 2–5 centimetres high, with evergreen leaflets arranged on opposite sides of a common stalk to form fan-shaped fronds up to 15 centimetres long. Because this fern has a small root it is able to exist in places where there is a limited amount of soil, and it is most likely to be encountered where the mortar on old walls is partly decayed between the bricks. Once it has become established, it is almost impossible to remove a specimen because the roots are long and wiry and travel a long way down. Interestingly this fern was once used as a herbal remedy for rickets.

On the wall you will also see some good specimens of one of our commonest native flowers, *herb-robert*, a member of the geranium family. This plant is easily recognised by its deeply divided five-lobed leaves, reddish-tinged hairy stems and five-petalled pink flowers, which have a habit of drooping down at night or in bad weather. If crushed between the fingers, herb-robert has a most unpleasant smell but it has many medicinal uses, for it contains tannin, an essential oil. It is used in herbalism to treat eye conditions, skin diseases, herpes and oral inflammations. What a delight to see the delicate flowers of herb-robert light up this old wall in high summer. Just before crossing the main road that runs between Polegate and Friston look out for a small patch of sandy soil beside the wall, for here you may see a group of *field digger wasps*, only about 13 millimetres long, with characteristic yellow and black bodies, they are busy making small burrows in which to lay their eggs. These small wasps, like most insects, have four main stages in their life cycle: egg, larva, pupa and adult, and are widespread in Britain.

The still air gently echoes with the cooo-coo, coo-coo, cu of a distant *wood pigeon* calling from tall ash trees in a small wood on the opposite side of this pleasant downland village. We make our way up Willingdon Lane to the base of the downs and then proceed up a track towards the summit of Combe Hill. It is pleasing to note that although much of this downland has been ploughed, there still remains some of the original turf, particularly about the raised embankments that are plentiful here. These are the lynchets of Celtic man, described in Chapter 1. On warm days in early spring the south-facing slopes are well covered with *gorse* in full bloom and scrub of *hawthorn* and *bramble*; also on the wing are early butterflies such as the *small tortoiseshell* and *brimstone*. Spring is late and remnants of winter snow still lie melting beside the path, their icy fragments glistening in the sunshine; but still you will find peeping through this last remnant of winter, the blue-purple flowers of the *early wood-violet*. The crinkly leaves of the *cowslip*, hugging the ground as if unwilling to show themselves, are evident everywhere heralding the glory of their flowers three or four weeks later.

The *cowslip* has declined greatly over recent years. Fifty years ago I remember many downland slopes around Eastbourne covered with their rich yellow flowers set against fresh green grass.

The ploughing up of our meadows, extensive use of herbicides and overgrazing have been the main cause of their decline, and here on the slopes of Combe Hill they will only show their true glory if cattle and sheep are kept off this undeniably rich pasture. The cowslip was an important plant to our ancestors as the leaves when young were often eaten in salads and the flower petals were used to make a wine which was said to be beneficial for giddiness and nervous disorders. The roots, too, are still used in homeopathy to treat pain and breathing problems and in the past they were used to cure insomnia and improve the memory. Truly a plant that has proved of benefit to mankind, and we must ensure its protection. Even as I write, a tractor is at work in the valley bottom, its dull roar breaking the serenity of this place. Gulls of many kinds compete with this man-made intrusion, their mewing and laughing calls piercing through the engine's roar as they fly incessantly behind the plough, only dropping down to jab at worms and grubs among the newly-turned earth.

Our path leads up towards the neolithic camp system and tumuli (burial mounds) near the summit of Combe Hill. In the early summer look out for the wild *columbine* plant on the north-facing slopes, which perhaps is a garden escape. It is 30–40 centimetres tall with drooping blue flowers, each of which has five petals. These petals have inward curving spurs arranged like doves around a food source giving rise to its English name, derived from the Latin *columba*, a dove. It is also known as aquilegia from its Latin name, which itself is derived from *aquila*, an eagle, because again the petals and spur resemble that bird of prey.

From April onwards to June, examine the soft mossy mounds that abound on downland slopes everywhere and are the nests of the *yellow hill ant*. Over a period of time the soil on these anthills becomes acidic, due to the ants squirting out formic acid as a form of defence. A lime-loving plant, the *rue-leaved saxifrage*, which can be found on walls and dry open spaces, seems to thrive on these anthills of the downs. Seldom more than 10 centimetres tall the rue-leaved saxifrage has three-lobed leaves usually tinged with red, and both leaves and stems have glandular hairs that secrete a gummy fluid that makes the plant feel sticky to the touch. The five-petalled pure white flowers are borne on long stalks in loose clusters and give a sense of neatness and beauty to this diminutive saxifrage.

We now look northwards over the Weald from the top of the downs here and the view is quite stupendous. As a boy, I lived in a bungalow at the base of Combe Hill and I remember climbing over great concrete blocks with iron rings embedded in them. These were the remains of the First World War Airship Station and the rings tethered those great ships to the ground. A few concrete bases with their rings can still be found in the woods at the base of the hill here, the remainder having been destroyed when the houses were built in the 1960s. The view from the top now has changed beyond recognition. Where once there were green fields and straggly hedges, or as Wordsworth so eloquently describes them, 'hedgerows, hardly hedgerows, little lines of sportive wood run wild', now there are lines of houses and street lamps everywhere. True, the distant Weald still looks much the same, but now perhaps there are more ugly modern barns and other buildings marring the landscape. Certainly, today the noise of not-so-distant traffic is a harsh reminder of man's destructive nature, and one feels that if allowed to continue unhindered, the very heart of our once tranquil and peaceful countryside will be lost forever. A *skylark* that arose quietly from grassland nearby now pours out its liquid song and brings you quickly back to the present. You remember the times past on the downs around Folkington and how numerous skylarks were then; and look across to those lovely sculptured hills and decide to visit them once again. Before leaving this hilltop with all its history, the words of that great poet Alfred, Lord Tennyson from *In Memoriam* seem to sum up all you have seen since leaving the Jevington churchyard just below:

> Now fades the last long streak of snow,
> Now burgeons every maze of quick
> About the flowering squares, and thick
> By ashen roots the violets blow.
>
> Now rings the woodland loud and long
> The distance takes a lovelier hue,
> And drown'd in yonder living blue
> The lark becomes a sightless song.
>
> Now dance the lights on lawn and lea,
> The flocks are whiter down the vale,
> And milkier every milky sail
> On winding stream or distant sea.
>
> Where now the seamew pipes, or dives
> In yonder greening gleam, and fly
> The happy birds, that change their sky
> To build and brood, that live their lives.

From land to land; and in my breast
Spring wakens too; and my regret
Becomes an April violet,
And buds and blossoms like the rest.

On the A27 just to the west of Polegate there is a signpost that reads 'Folkington Lane'. This lane has a quality and expectation that appeals, so walk up it on a cold clear day in late winter. On turning the first corner on the road you see through the naked branches of tall trees a huge panorama of rolling downland, truly a most magnificent sight.

A small group of scattered houses along the lane make up the village of Folkington; since its Saxon name 'tun' or 'ton' means farm it could be any little downland village, nestling among the folds of these chalk hills sheltered from south-westerly gales. As we shall see, its particular flora and fauna is repeated many times over in East and West Sussex. The Church of St Peter, set in a delightful place just below the wood at the end of the lane, is typical of this rural county. Dating back to the thirteenth Century, the church is a small building built of stone, faced with flint in the Early English style. In March and April the grassy bank in front of the entrance is a blaze of yellow from masses of *primroses* which grow all over it. The hum of bees in warm spring sunshine brings another dimension to this place and adds to the fascination and charm of Folkington.

Saunter up into the wood through ivy-covered trees and let the crisp brown leaves of beech rustle and swish beneath your feet. On sheltered banks lie *snowdrops*; their flowers etched with delicate grooves of green, and their lovely forms give the wanderer through the wood an uplift of joy in the still dark days of late winter. Spring is not far away and the snowdrop flowers are full of rich nectar to attract the first insects out from their hibernation to do their work of pollination. It is hard to imagine that later in early summer this same somewhat barren place is covered in green foliage of many kinds with bluebells and wild garlic adding colour and pungency to the scene. Another of our fifty species of wild orchid still grows here, the *white helleborine*. About 30 centimetres in height, it is conspicuous with creamy-white flowers and fresh green leaves which stand out against the carpet of rich-brown dead beech litter. Their fibrous roots penetrate deeply into the lifeless leaves and soil to as much as 40 centimetres. The flowers which appear in May are much frequented by wasps which have no tongue and find the nectar easily accessible, lying as it does in a hollow in the lower lip of the flower. The seed of this orchid may take eight years to germinate and a further three years before flowers appear. Fifty years ago, I remember the white helleborine growing in profusion in Folkington Wood but now you will be lucky to find just a few of them.

An inconspicuous little spring flower that seems to flourish in the woods and hedgebanks around Folkington and Jevington is *moschatel*. Some 5–10 centimetres tall, this plant spreads over the ground by an extensive underground root system. The greenish-yellow flowers, which appear in March, are held at the end of the stems and each flower head contains five individual blooms, four of which are arranged like the faces of a clock; hence another name for the plant, 'townhall clock', while the fifth flower points upwards to the sky. Each stem-leaf is cut into three leaflets; but there are also leaves that spring direct from the rootstock and these may be cut into one or two sets of three leaflets. In damp weather, the flowers give off a faint smell of musk. The Latin name for the plant, *adoxa* is derived from the Greek privative, *a-*, and *doxa*, opinion, which signifies that the plant is of no account. However, it is worth searching for just to admire the symmetry of the flowers.

As you emerge from the west side of the wood a magnificent sight greets you; a sweeping vista of downland where in the evening light dark shadows from a low sun hang tentatively on steep slopes. Between the slopes are hollows where a good depth of soil allows tall trees of *ash* and *beech* to flourish and form small rounded woods. From their midst and across fields of winter-sown wheat comes the strident and far-carrying call 'kok-kok-kok' of the *pheasant*. *Rooks*, too, call out their raucous 'kay...kaw...kaw' as they busy themselves with nest building.

The *rook* is easily recognised from other members of the crow family by the grey-white warty skin around the base of the beak; also when walking its thighs appear noticeably shaggy. A gregarious bird, it nests in sometimes large colonies of over 1,500 pairs. Their large nests are built among the highest branches of tall trees. Nest building begins early in January, the nests consisting of twigs and sticks knitted together with mud and clay, lined with straw, hay or wool. The four or five pale green eggs blotched with smoky-brown are laid from February to May, the laying season varying according to the weather conditions. Sometimes the rook will nest in chimney pots, and this has become more common since the 1960s as coal fires became unnecessary and only warm air ascends from redundant fireplaces. People are now more in the habit of feeding birds by throwing waste food onto their back lawns and so, like the gull, the rook is becoming a suburban bird as well as a bird of our countryside.

In Sussex the rook population has declined by almost half in the last 50 years, probably due to the reduced availability of their traditional food, particularly grubs, insects, worms and grain. These have all been affected by changes in agricultural practice where the use of pesticides has reduced significantly the supply of invertebrate foods. Changes from spring to autumn sowing of cereal crops has also depressed the supply of grain during the breeding season. However, it appears that the rook population has now stabilised and certainly my own observations at East Dean, near Eastbourne, suggest that the population is increasing; the number of nests in the local rookery has remained at about fifty since the 1950s, when I remember climbing up to them as a young boy. Of course there have been fluctuations in the numbers during intervening years. Interestingly the total population around the village has increased recently due to their being given a plentiful supply of food by friendly people in their gardens. Sadly this has had the effect of reducing the population of garden birds because the rook, like many of its family, is a rapacious killer.

Outside the breeding season rooks gather in large numbers in fields, to feed off any plentiful supply of food. At sunset they meet other flocks and assemble together in their favourite wood to roost. This is particularly the case in winter when often they roost in the company of *jackdaws*. They are both light sleepers and at the first hint of daybreak they are up and off to their feeding grounds in nearby fields. Many believe that rooks are beneficial to agriculture because they take out harmful grubs such as wire-worms, but studies have shown that on balance the opposite is the case and the destruction of newly planted corn is particularly damaging. Not only do they damage crops, but the bird is also a danger to garden and ground-nesting birds and can often be seen quartering fields and hedgerows to steal and eat the eggs and young of anything it finds. Young *robins* in my own garden are taken almost every year by rooks. At one time there was some control on their population and young rooks were removed from their nests, then skinned and soaked in milk for twenty-four hours before cooking as rook pie. This pastime has long since ceased and the rook is left with virtually no predator to control its numbers, except man himself.

Following the path that skirts the southern edge of the wood you soon come to the base of the Downs themselves. Nearby, amongst scattered shrubs of *wayfaring tree*, *bramble* and *hawthorn*, you may be lucky to find a very rare species of honeysuckle, the *fly honeysuckle*. This is known elsewhere in Sussex only on the north-facing slope of the Downs just to the south-east of Amberley. It differs from the ubiquitous honeysuckle by having smaller, 1-centi-metre long yellow flowers which grow in pairs at the end of slender stems; its leaves being more oval-shaped and covered in fine hairs, particularly underneath. The fly honeysuckle is about 2–3 metres high and flowers in May, generally a month earlier than its better known relative. Here at Folkington it is situated close to an ancient trackway, as it is in West Sussex; perhaps it was brought here by early man from continental Europe where it is widely distributed.

The downland escarpment is characterised by short, dense, well-drained springy turf which has a great richness of flora. In May and June you will find masses of *bird's-foot trefoil*, *salad burnet*, *wild thyme*, *common rock-rose* and many species of orchid including the *bee orchid*. Butterflies, too, abound on these flower-strewn slopes and amongst others, you will see the *small tortoiseshell*, *gate-keeper*, *meadow brown* and *common blue*, all on the wing in June and July, their colours adding great beauty to an already brilliant scene.

Ascend the steep path that leads on to Alfriston. On your right is Hunter's Burgh, the most easterly of the Sussex long barrows, where neolithic man buried his dead, usually laid at one end of the higher and wider side which faces east. So on and upwards to the prehistoric trackway which traverses the ridgeline of this lovely stretch of downs. There is a great sense of freedom and space here and the sky seems limitless. The views stretch over ripening corn-fields and over eye-catching areas of vivid yellow *oilseed rape* crops. To the south, across the ever-changing colours of Friston Forest and Lullington Heath, is the sea, a sparkling blue; to the west lies far off Firle Beacon, a magnificent hill that dominates the northern escarpment of the South Downs and forms the centre-point of the next chapter.

Walk along the ridgeway that follows the rim of these downs and enjoy the atmosphere of uncrowded hills and feel a sense of timelessness, even if for just a short while. *Skylarks* sing loud here in the clear air. Watch them as they hang on fluttering wings and pour out their liquid song and see them gently descend to their nests in the deep long grass of summer. At the top of Folkington Hill are more burial mounds of early man, in the form of raised banks of earth called tumuli and marked on maps as such. In the near distance are the stark ruins of Hill Barn, standing grotesquely amongst all this beauty; this must have been a working farmstead at one time as there is evidence of a walled garden, fruit trees and a place for storing water collected from the roof. What interesting stories these remains could tell of smugglers, highwaymen and just ordinary farming folk!

As you descend the easternmost ridge look across the valley to Wannock, the name of which in old English appears as *weala hnoc* or 'nook of the foreigner'. When the Saxon raids were at their peak in the fifth century AD, the last of the Romano-British took refuge and hid in Wannock Glen where its stream, and thick cover of trees and shrubs gave sustenance, concealment and shelter. After the Saxons dominated and controlled the surrounding countryside these remaining British were known as 'the foreigners'. Eventually of course they were subdued, killed or absorbed into the Anglo-Saxon culture.

At the bottom of the ridge, there is a sunken trackway dating back to prehistoric times and used extensively by smugglers in the eighteenth and nineteenth centuries; it leads back to Folkington Church from Birling Gap, Crowlink and Jevington. Beside this track there are some age-long earthworks on which rabbits have browsed for centuries and kept the grass very short. In June and July look here for the diminutive green and brown *frog orchid* which is an uncommon species, and its occurrence in any locality is spasmodic, some years being plentiful and others entirely absent. Many people are disappointed when they find it for the first time and look hard for a resemblance to the four-legged amphibian. If a frontal view of the flower is taken, a sort of likeness can be seen to the head of a frog with the lobed lip representing its out-thrust tongue.

In this area on private land you may just stumble on a very rare and beautiful plant, the *rough marsh mallow*. It was discovered at this site only a few years ago and is now increasing in numbers. The rough marsh mallow blooms in June and July and seems to prefer disturbed ground on chalky soils. There is some hope that it will spread from its original locality and become locally common in the area. It can be recognised by its solitary five-petalled pink flowers all of which have fine dark pink veins running through them. The whole plant rarely exceeds 15 centimetres in height and is covered in fine shiny hairs looking almost as if someone had sprinkled droplets of ice all over it.

As you walk these paths over the Downs in midsummer you cannot fail to spot the *common poppy*, splashing red the edges of cornfields. Where 40 years ago whole fields were ablaze with red, now they, with other cornfield weeds, are confined to the edges if not taken out with modern herbicides. It is important that farmers realise the importance of retaining as much as possible of our field-weeds for their seeds are life-sustaining for many birds such as the *corn bunting* and *skylark*. So, wherever possible, the edges of our fields, if only for a metre or so, should be spared the destruction of the crop-sprayer.

Before following the trackway on to Wilmington it is worth reminding the reader of nearby Lullington Heath and Deep Dene just over on the other side of the downs. The natural history of these areas with their *adders, nightjars, adonis blue* and *silver-spotted skipper* butterflies to name but a few of their specialities, was fully covered in my book *A Natural History of the Cuckmere Valley*. I will not repeat what was written there but I will just add details of four interesting species that I did not describe then.

First of all, the *dormouse*, which is now present in quite large numbers on the Lullington Heath Nature Reserve and in nearby Friston Forest. Dormice can be recognised by their plump 8-centimetre long chestnut-coloured body, small ears, large eyes and long bushy tails. They look like small squirrels rather than mice and were often kept as pets by children in the Victorian era, but they have now become a rare species. The key reason for their decline is, as usual, habitat destruction. Their ideal home is *hazel* coppice that is cut regularly but not too frequently, so that it grows large enough to produce good crops of nuts. If the coppice is neglected then other trees will overgrow the canopy and shade out the bush layers that provide the dormice with their food. Here on Lullington Heath they seem to thrive in the thick undergrowth and feed not only on hazel but also on *beech* nuts; one sign of their presence being a neatly gnawed hole in the nut.

The dormouse is more active at night, when it is agile and lively, climbing well over bushes and trees, looking for berries, nuts and small insects. It constructs its winter nest below ground or in a hollow tree stump where it hibernates from September to April curled up in a tight ball, chin resting on its belly and its tail curved up over its head. The summer nest is built as a loose ball of grass, leaves and shredded honeysuckle bark, and is usually wedged in the branches of a bush or small tree. The female dormouse produces one or two litters per year, each of three to five young. The lifespan of this small mammal is about four years, relatively long, due to the fact that it spends a good deal of its time asleep out of reach of its enemies. As on many nature reserves the dormouse is encouraged by the placing of nest boxes specially constructed for it and strategically placed in suitable locations.

In Wilmington churchyard below the Downs, grows an interesting tree, a *yew* that is reputed to be about sixteen hundred years old. It is some 15 metres tall with a massive base from which, slanting at opposite angles, grow two hollow boles which are probably the halves of a single stem that split many centuries ago and have now healed.

There are at least 500 churchyards in England and Wales each containing yew trees which are as old as the churches themselves, and in many cases probably a good deal older. Many theories have been put forward as to why, in the western world at least, yews of great age seem to coexist almost exclusively with places of worship. One theory has it that they were planted in protected churchyards to provide wood for longbows. This can be discounted because the English yew was too brittle for use in longbows and the wood was usually imported from Spain and Italy. Other theories suggested that they were planted to give protection and shelter to congregations gathering before the church doors were opened or to provide decoration for the church. However the most probable reason is that the trees' longevity, their funereal appearance and their association with ancient religions such as those of the Druids and Celts meant that they were identified with mortality and the life hereafter. Certainly many of the churchyard yews are older than Christianity itself and there is evidence that they were planted on top of graves. Interestingly when the ancient elm of St Mary's churchyard at Selborne in Hampshire fell down in the great gale of 25th January 1990, excavations of its root system revealed 30 skeletons, some going back to about AD 1200. It seems certain that even before Christianity the yew was a sacred tree and it was pragmatically sanctified and accepted by the Christian church. From mediaeval times it became the practice to plant two yews in each churchyard, where during a funeral the coffin would pass beneath them.

The yew is a species that thrives on well-drained chalk and limestone soils and so it is common on the South Downs, where it flourishes in ancient woods such as Kingley Vale near Chichester, perhaps the finest site for it in Europe. In winter this dark and solemn tree is very prominent, surrounded as it usually is with more abundant and leafless trees such as the *beech*, *sycamore* and *ash*. The yew keeps every leaf for several years and only discards a few at a time. Every tree is either male or female and in February the male trees are covered in catkins, which if shaken give off puffs of yellow pollen. These are wind-borne and fertilise the more minute female flowers, hidden in scales under the leaves of a different tree. By October the flowers have become berries, each of which is made up of a single hard seed surrounded by a collar of bright scarlet and waxy pulp. These are much favoured by birds which spread the seed widely. The wood of yew is rust-red in colour and is hard, strong and very durable; it is therefore much sought-after for woodcarvings, fine furniture and turned bowls and plates. Let us leave this ancient tree of Wilmington with the words of Lord Tennyson again in his *In Memoriam*; as a boy he must have been familiar with the yew on the chalk hills of the Lincolnshire Wolds.

> Old Yew, which graspest at the stones
> That name the under-lying dead,
> Thy fibres net the dreamless head,
> Thy roofs are wrapt about the bones.
>
> The seasons bring the flower again
> And bring the firstling to the flock;
> And in the dusk of thee, the clock
> Beats out the little lives of men.
>
> O not for thee the glow, the bloom,
> Who changest not in any gale,
> Nor branding summer suns avail
> To touch thy thousand years of gloom.
>
> And gazing on thee sullen tree,
> Sick for thy stubborn hardihood,
> I seem to fail from out my blood
> And grow incorporate into thee.

Before leaving the churchyard look at the bright golden star-shaped flowers of the *lesser celandine* plant which bring so much colour to our countryside in early spring. The lesser celandine has a fibrous rootstock which produces a large number of cylindrical tubers which readily break free and produce new plants the following year. These tubers were said to resemble haemorrhoids and in the seventeenth and eighteenth centuries herbalists believed in the 'doctrine of signatures' which proclaimed that God had given to every plant a physical clue as to its medicinal qualities; so the lesser celandine was used then to treat piles and it was known as pilewort. Interestingly, lesser celandine is common in churchyards everywhere and this gave the belief extra credence because herbal plants that grew in sanctified ground were considered to be especially powerful. Perhaps that is why there is such a great variety of similar plants in our churchyards; such places are now considered to be valuable sites protecting and preserving our flora from the destructive nature of man.

The fourth interesting species can be found beside the track that leads across the old Wilmington Airfield just to the east of Sherman Bridge. In June and July look for the bright yellow (sometimes cream-coloured) flowers of the *moth mullein*. The plants are up to a metre tall and can be distinguished from other mullein species by being almost devoid of hairs, having a slightly angled stem, and by having solitary flowers held at the end of short stems. Another distinguishing feature is that the anthers in the centre of the flower are also purple in colour, a characteristic it shares only with the *dark mullein*. The moth mullein is a rapidly declining species and now must be considered as a rare plant of our waste places. This elegant plant has existed on this particular piece of waste ground for many years and it is to be hoped that any development of the land here does not destroy it. See it swaying gently in the breeze, and shining out brightly in the late evening just as the sun sets over the distant escarpment of the South Downs. What a beautiful plant in a setting all of its own.

Before starting on the next chapter return with me in late summer to the Downs that dominate Folkington Church and its surrounding cluster of houses. The fields of wheat are a rich golden colour which change hue as the play of the wind ripples through their lush growth. A flock of *goldfinches*, with their bold markings of red, yellow and black, chatter as they feed on thistle heads that abound around the field edges. The sky changes to a deep orange glow as the sun sinks behind the tall hills. Rest awhile on the grassy bank beside the path near the wood and ponder at the sight of these sloping Downs, their hollows fast filling with dark shadows. Listen to the silence of the scene and let the spirit of this place give you a peace that is so deep and calm.

Inexorably the earth tilts on its axis; the first beech leaves are starting to change colour to a golden-brown; autumn is nibbling at the trailing edge of summer. Time to move on to more distant hills.

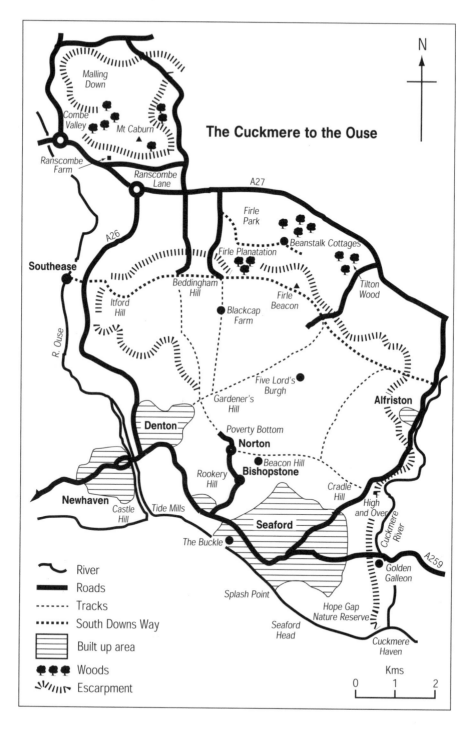

The Cuckmere to the Ouse

River
Roads
Tracks
South Downs Way
Built up area
Woods
Escarpment

N

Malling Down
Combe Valley
Mt Caburn
Ranscombe Farm
Ranscombe Lane
A27
Firle Park
Beanstalk Cottages
Firle Planatation
A26
Southease
Beddingham Hill
Itford Hill
Firle Beacon
Tilton Wood
Blackcap Farm
R. Ouse
Five Lord's Burgh
Alfriston
Gardener's Hill
Denton
Poverty Bottom
Norton
Beacon Hill
Rookery Hill
Bishopstone
Cradle Hill
Newhaven
Castle Hill
Tide Mills
High and Over
Cuckmere River
A259
Seaford
The Buckle
Golden Galleon
Splash Point
Hope Gap Nature Reserve
Seaford Head
Cuckmere Haven

Kms
0   1   2

4

THE CUCKMERE TO THE OUSE

Mount Caburn

The South Downs between the two rivers, the Cuckmere and the Ouse, are the least inhabited and are endowed with wide open stretches and few trees; the long valleys and folds gently touch and embrace each other to give a feeling of remoteness that is real, despite the cornfields, cattle and grazing sheep. Here is a place where no busy road spoils your sense of solitude and few farmsteads mar your view of settled permanence, and where cumulus clouds form their cottonwool-like heads as the sun warms up the air over the hills on a summer's day. The most prominent hilltop on these spacious downs is Firle Beacon; at an elevation of 217 metres it is the highest point in East Sussex and the view from its summit is unimpeded for many miles in every direction. Across to the north-west is the small range of hills set apart from the main escarpment, and overshadowed by the magnificent crown of Mount Caburn; this is a unique area of interest which we will explore later at the end of the chapter.

We start in the region just south of Alfriston at a place called High and Over. Here, over millions of years the Cuckmere has eroded away the underlying chalk to form a steeply sloping escarpment now almost completely covered by scrub of gnarled *hawthorn*, *elder* and *wild cherry*. In April watch out for the arrival of a small but active and very beautiful falcon, the *hobby*. It is an astonishing flier as it swoops effortlessly in graceful arcs, catching flying insects on the wing, especially at dusk. Enjoy the wonderful views from the top of High and Over; the Cuckmere below a gleaming ribbon of blue set upon a carpet of green as it meanders gently southwards towards a distant and sparkling sea. Dominating the western skyline, the trees of Friston Forest with their ever-changing colours; a purplish hue of beech buds in spring, through the deep green of summer, to the orange and browns of autumn. Do not leave this area before exploring the wonderful downland slopes of nearby Cradle Hill with their orchids and butterflies. The gentle walks here through young woods of ash and sycamore inter-mingled with hawthorn and blackthorn scrub will reveal birds like the *long-tailed tit*, *wood pigeon* and the occasional *pheasant*.

The professional as well as the amateur naturalist will not wish to pass by the wonderful area of Seaford Head and the Hope Gap Nature Reserve; there is so much to see here that I can but just touch the surface of what it holds.

Seaford Head and its surrounding area is much visited by people walking their dogs and one could suppose that such disturbance would inhibit the bird-life of the area; not a bit of it. For here you may see the *peregrine falcon*, *shelduck*, *wheatear*, *kittiwake*, *fulmar*, *skylark*, *meadow pipit* as well as many *gulls* and a large variety of *warblers*.

One bright sunny day in late April I heard the loud chatter 'hec hec hec' of a bird which I suspected to be a female *peregrine*, or falcon as she is called. Cautiously I crept over the brow of a small hill and then I saw her perched nervously on a great pinnacle of chalk just thirty metres away. The falcon is of similar appearance to the male, known as the tiercel, but is much larger and darker; both have a black crown and bluish-ash upper parts which contrast with their buffish-white underparts barred with dark brown and grey. As I crept slowly forward to get a better view she tensed up, put her head forward and flew low, then soared upwards and inland. What a sight.

The history of this falcon on the great cliffs of Sussex is interest-ing. In the early part of the last century there were seven to twelve breeding pairs occupying eyries on their ancestral cliffs. In 1940 the then Air Ministry organised the destruction of peregrines by shooting, in an attempt to protect military carrier pigeons which carried important messages from agents in occupied France. The policy was successful and their destruction was almost complete in Southern England and Wales, although many continued to thrive in the north of England and in Scotland. Sussex falcons seemed to have fared better and six or seven pairs continued to breed throughout the war. However, by 1946 the number of breeding pairs dropped to four. They hung on precariously with numbers fluctuating between three and six pairs until May 1956, when the last breeding pair was recorded at Seaford Head. The reason for their decline since the war years is almost certainly due to the use of pesticides and other toxic chemicals on farms to control harmful insects and weeds. Persecution by gamekeepers, egg collectors and pigeon fanciers also played a part, but it was the use of chemicals which entered the food chain and poisoned the peregrine, that caused the rapid decline of this magnificent bird of prey. Many other birds such as the *sparrowhawk* were similarly affected in the 1950s and 1960s, and as a result between 1962 and 1969 even tighter controls were imposed by the government on the use of harmful sprays and seed dressings. In 1967 the peregrine population throughout Britain started a slow recovery, and in 1990 the falcon was breeding again at Beachy Head. Now in the new millennium we have perhaps ten or more pairs breeding in Sussex not only on the cliffs but also on tall buildings in Brighton and inland near Lewes.

The peregrine nests on cliff ledges where it makes a small hollow in the debris and occasionally lines it with pieces of seaweed, hair and small sticks. With the increase in the feral pigeon population of our towns and cities, the peregrine, with an eye on a continuous and easy source of food, has now started to nest on ledges of tall buildings in these places. It is becoming a suburban and even an urban bird; what a success story! The three or four eggs which make up the clutch of the falcon are laid in late March and early April. They are coloured a rich buff and red with grey blotches and unlike many raptors' eggs their incubation does not begin until the last egg is laid. This ensures that all the eggs hatch out nearly simultaneously after about 31 days. Young peregrines are covered in white down at first, but feathers start to show after about 18 days when the characteristic moustache-like streak below the eye begins to appear as a brown patch. As the feathers grow, so the young begin to practise wing-flapping before leaving the nest after about 38 days. During this time the tiercel catches most of the prey for both the young and his mate, although when the young are about 20 days old the falcon is able to leave the young and hunt for her own food. What a sight it is to see the peregrine dive down vertically at high speed, estimated at over 200 kilometres per hour, onto an unsuspecting bird like a pigeon, gull, or blackbird, and then to swirl upwards again on sickle-shaped wings. The peregrine hits its prey with great force with its extended claws and carries it to a perch, and only a bunch of feathers is left fluttering down to the ground to mark the point of impact and almost instant death.

After flight the young peregrines are still dependent on their parents for food but gradually they learn how to catch and kill prey for themselves; by the autumn they are fully independent and leave their parents to find new territories, sometimes travelling long distances and even to other countries such as Iceland.

Observations suggest that peregrines seldom live for more than 10 years, so let us hope that they continue to nest undisturbed and unpersecuted on our great chalk cliffs and their breeding is successful enough to increase their population and give pleasure to people for generations to come.

Just by chance as I was stalking the falcon I happened to notice a large rabbit hole with pieces of fluffy down around the entrance, a typical sign of a *shelduck*'s nest site. Just as I was trying to remove some of the down to make the position less conspicuous, a pair of these birds appeared above me circling with slow wing-beats, the broad chestnut bands around their bodies easily distinguishing them from other ducks. Clearly the drake shelduck with its distinguishing prominent red knob on its bill, had just returned with his mate after leading her off to the lower Cuckmere marshes for a feeding session. Shelducks particularly like rabbit holes in which to build their nests, using grass and down from their own bodies. Occasionally the shelduck will dig its own burrow or adopt that of a *fox* or *badger*; and sometimes the nest is as deep as 3 metres or more in the hole. The ten to twelve cream-coloured eggs are laid from April to June and the attractive little white ducklings, spotted with black, emerge some five or six weeks later. Both parents then lead the young to the shelter of nearby water, in this case probably the Cuckmere river. The ducklings grow rapidly and in summer they group themselves with other families into 'crèches', for at this time most of the adult birds leave for their moulting grounds on the Continent, leaving just a few parents to look after the young. After moulting and growing new feathers, the adults begin to return to their winter feeding grounds on the south coast from October onwards. By this time, however, the young from that season, now truly independent, have grown restless and start their own migration east and west along the Channel coast and even as far as mainland Europe. The shelduck is a common breeder in Sussex and Hampshire, particularly around Rye and Chichester Harbour, but several breed inland, particularly in the Arun valley. Most are resident over winter but when particularly severe weather occurs the birds leave to fly south.

A word of caution as you walk the grass-covered slopes near to the cliffs at Seaford Head, for this is a favourite place for basking *adders*, particularly on a warm spring day. The adder is Britain's only venomous snake; its venom is a powerful heart depressant, causing rapid death of its natural prey. If left alone this snake will not attack humans, and in any case its bite is seldom fatal and only twelve people have been killed in Britain by them in the last sixty years. Be careful and treat them with respect; if about to be trodden on they may rear up and hiss at you giving you some warning of attack. Adders have a distinctive dark zigzag along the back and a V or X mark behind the head. The background colour of this snake shows wide variation between whitish or pale grey through pale yellow to brown; generally the female is a duller colour and is larger – up to 61 centimetres long.

The first adders to be seen in March and April are the males just emerging from their 5-month-long hibernation in a hole in the ground. About a month later the females follow, taking up station themselves in territories not occupied by males. At this time adders are lethargic, do not feed, and are content to lie about in the sun. They do not usually leave their cover until about 10 a.m. retiring again by mid-afternoon. By the middle of April the males will have shed their skin, to become brilliant in colour and then begin an active and bad-tempered search for females. This is the time that people and their pets are most often bitten by adders. Rival males will often fight by rearing up from the ground, ducking and weaving with their fore-parts as if fencing each other. No biting actually occurs and eventually the loser is chased away, the winner being left with the female. After mating the female spends more time basking in the sun and slowly swelling as her eggs develop. Young adders are born some three and a half months after conception. The average number born to each female varies from 5 to 20 and the young adders normally feed and enter hibernation with the adult. The adder's diet consists chiefly of lizards, slow-worms and small mammals, the scent trail of their prey being followed with the snake's tongue, which is its organ of smell. Occasionally young birds in their nest are taken; I remember once seeing *whitethroat* nestlings being swallowed by a large adult adder which had climbed up nearly a metre through the branches of a small bush to the nest.

Another bird that nests in the numerous rabbit holes around Seaford Head is the *wheatear*. This delightful bird, distinctive with its white rump, black tail and blue-grey back and head, is one of the first of our summer visitors, arriving from its winter quarters in Central Africa in March or even late February. It is conspicuous as it flits from one grassy mound or chalk outcrop to another. The wheatear used to be one of the typical summer birds of the chalk downlands only fifty or sixty years ago, but now only a small number of pairs are confined to just a few coastal sites and the occasional one inland. In April and May the wheatear lays her five or six unspotted pale blue eggs in a nest made of dead grass and moss, lined with wool, hair and rabbit fur, clumsily constructed at the end of a rabbit burrow or other hole in the ground. Occasionally it will nest under a boulder or slab, and I remember that on the Eastbourne Crumbles several pairs used to nest under old rusty pieces of corrugated iron. I understand they still nest in such places in nature reserves at Rye Harbour and Dungeness. Their presence on our nature reserves is a good sign, for they are susceptible to disturbance and if we are to retain them on our chalk downlands we must give them places where they have a chance to breed successfully. At one time in the eighteenth and nineteenth centuries, shepherds supplemented their wages by catching wheatears with nets. They were then sold for two old pence a brace to the markets in Brighton and London, where they were bought as a great delicacy for the dinner tables of the gentry.

Occasionally you may see a *rock pipit* looking for nest material or insects amongst the grass of the cliff-top, but more usually this bird will be seen amongst the chalk boulders and seaweed-covered rocks of the shore. Like many of the pipits the rock pipit is somewhat dull in appearance with greyish upper parts and whitish underparts which are streaked with brown in the autumn and winter; these streaks being absent in the breeding season when the bird takes on a flushed pinkish colour.

The rock pipit is slightly larger than the *meadow pipit* with a length of some 16 centimetres as opposed to 14 centimetres, and its legs are much darker than the latter species. However, you will not go far wrong if you treat those pipit-like birds on the seashore in the breeding season as rock pipits, and those of the open pastures and fields as meadow pipits. The rock pipit is not a common bird in Sussex, but some ten or so pairs nest along the cliffs between Brighton and Eastbourne, and also near Hastings. The nest is built of grass and seaweed lined with hair and feathers in a crevice in the chalk cliffs. The four or five greyish-green eggs, closely spotted and mottled with reddish-brown, are laid from April to July. Enjoy watching this bird as it runs in and out of the water, amongst the waves, or hops about weed-draped rocks searching for minute scraps of marine life.

Before leaving Seaford Head and its associated cliffs, it is worth a walk down towards the town itself. Here, at a place called Splash Point, four or five pairs of *kittiwakes* started to nest in 1976, and since then their numbers have increased, fluctuating between 50 to 300 pairs. They nest in colonies on cliff ledges and once the site becomes overcrowded many pairs leave to form new colonies nearby. This delightful gull with its dove-like appearance, black legs, and black wing tips, gets its name from the loud plaintive call of 'kitti-wa-ak' or 'kaka-week' which it utters around its breeding stations. Like the *fulmar petrel*, the kittiwake is a truly oceanic species, spending much of its time amongst the waves of the deep Atlantic, coming to our coasts in February and March to breed. What an attractive bird to watch at this time, as they pair up, standing close together, breasts inwards and touching the cliff face and each other. Often they break out into soft mewing cries which spread to the whole colony. Then suddenly all is silent and only the splash of waves against the promenade and the gentle wash of sea on pebble shore can be heard. Their courtship exchanges are delightful; graceful bowing of heads to each other, and gentle fondling movements with their bills, all accompanied by excited mewing calls.

Kittiwakes make substantial nests of grass and seaweed and their two or three eggs are laid in May and June; both parents sharing the incubation of the eggs, and after about 22 days the young hatch out. In contrast to the somewhat savage and often canniba-listic behaviour of some of the larger gulls, the kittiwake never preys on the eggs or young of its own kind or that of any other species and is truly an inoffensive and likeable bird. It feeds off small fish, plankton and crustaceans taken from on or near the surface of the sea. The young are fed on this food for about five or six weeks after which they leave the nest and fly out to sea.

By the end of August the breeding ledges are empty but with luck they will be occupied again early the following year. Mean-while, wish the kittiwakes well as they leave this friendly Sussex shore to ride out the autumn and winter storms amidst the flashing foam of mid-ocean.

The kittiwake as a gull could easily be the bird described in Sir John Squire's poem 'The Birds' where he wrote:

A dizzying tangle of gulls were floating and flying,
Wheeling and crossing and darting, crying and crying,
Circling and crying, over and over and over,
Crying with swoop and hover and fall and recover.

For a complete contrast to the cliff ledges, walk down the gentle slope eastwards to the scrub and thickets that comprise much of the Hope Gap Nature Reserve. Here in early summer you will hear and occasionally see many of our migrant warblers like the *willow warbler, chiffchaff, whitethroat, blackcap, garden warbler*, or even *grasshopper warbler*, to name but a few. Many of these nest among the abundant cover of the reserve, except for the last-named, which although once relatively common, is now almost extinct in Sussex.

This area of East Sussex is one of the finest to watch the spring and autumn migration of birds. Sit on the grassy slopes in early spring when a southerly wind brings in the first migrants such as the *wheatear, sand martin, swallow* and the *cuckoo* to be followed later by the *osprey, swift* and the *warblers*. You may be lucky to catch sight of a rarity such as the *hoopoe, bee-eater* or *red kite*. For many thousands of years our birds and their ancestors have undertaken their many and varied journeys over sea, land and desert. Clearly, breeding and food motivates them to a great extent, but nobody really knows why so many birds migrate. What great mystery of evolution compels a bird like the *Arctic tern* to fly halfway around the world and back in a year? How do the birds navigate over the great oceans, at times through fog and great storms?

Perhaps one explanation could be associated with the phenomenon of continental drift. Some 300 to 400 million years ago the continents were joined together in one great land mass. Gradually this land mass split apart and huge pieces drifted away from each other on plates within the earth's crust, so forming continents. Over the past 200 million years the continents slowly established their present positions, although even now they are still shifting relative to one another. At one time, then, the ancestors of our birds, the reptiles, lived in one place, relatively speaking, and had no need to travel great distances for food. Imperceptibly, over the aeons of time, the first flying animals (birds) of the early Cretaceous period 120 million years ago needed to travel further and further to get to their food and this distance got greater and greater as the land masses drifted apart. Their breeding cycle, too, would have been affected, and so over many millennia birds gradually evolved the skills necessary to navigate the increasing distances required to meet their needs. Whatever the reason, the birds give us a marvellous spectacle as you see them congregating sometimes in large numbers, ready to leave on their long flights.

Seaford Head and its associated nature reserve has an abundance of wild flowers. Chief amongst these is the *sea pink* or *thrift* which adorns the brown and white cliffs with its delicate rosy-pink flowers in late April and May, a common enough plant all around our coasts, although in Sussex it is somewhat local, but where it does grow it is sometimes abundant. May is the best month to see *thrift*, when its masses of pink blooms contrast so exquisitely with the blues of the sea and sky. Long after the flowers have withered and died in late autumn their outer remains survive as papery sheaves. These clusters of pale brown give a certain quality to the bleak landscape in midwinter, contrasting well with the blue-green cushions of the living plant itself.

Along the top of the cliff near Hope Gap is a large patch of that very poisonous plant of the nightshade family, *henbane*. With stout stems nearly a metre tall, this plant has an unpleasant smell with hairy, sticky, light-green leaves. Its drooping yellow and purple flowers bloom from June to August. During the fruiting stage the seed-heads resemble a row of molars and for this reason it was once considered a cure for toothache. The plant contains large amounts of the hallucinatory and poisonous drugs hyoscyamine and hyoscine, which are now used as a basis for painkillers. One can imagine the effect on patients being treated with it for tooth-ache! Henbane flourishes where the soil has been disturbed, such as around rabbit warrens, so look for it in such places and near the cliff edge itself where sods of earth have been torn up by winter gales. You will enjoy finding this interesting plant but treat it with respect when you see it.

Another plant to be found growing amongst the tufts of short grass and low growing bramble around Hope Bottom is the *gladdon*, or *gladwyn*, an old English word for a sword. A member of the iris family, it has short-lived, purple-violet, darkly veined flowers which bloom from May to July. Gladdon is easily recognised not only from its flowers but also by its distinctive, long dark lance-shaped leaves, which turn brown and yellow at the tips, scorched by the salt-laden winds. The seeds of this plant when fully ripened are a brilliant orange and lie in rows like peas in a pod on the open segments of the seed capsule all winter. For this reason they are frequently used to adorn graves in the churchyard in winter months and bird-sown seeds cause the plant to become naturalised beside nearby hedges. If crushed between the fingers, the leaves of gladdon give off an unpleasant odour of stale beef, hence its other names, *stinking iris* and *roast-beef plant*.

In May and June look for the small four-petalled white flowers of the *field pepperwort* which grows in many places amongst the chalk grassland here. The triangular toothed upper leaves of this plant clasp the stem while the lower untoothed leaves form a loose rosette that soon withers. The roots and leaves of pepperwort are bitter tasting but the leaves when dried and powdered were once used as pepper. The group of pepperwort species are all somewhat similar and can be distinguished from each other by the shape and structure of their seed pods, differences which are perhaps a little too detailed to go into in a book such as this.

Before leaving the wonderful nature reserve with its many other treasures take one last look at the rhythmic grandeur of the Seven Sisters on a bright sunny day when puffy white clouds play above their scalloped edge and a shimmering sea adds to the atmosphere and contentment of the place. Pause awhile and feast your eyes on this sight, said to be one of the finest in England.

Briefly we move across the main A259 road having refreshed ourselves at the excellent restaurant of the Golden Galleon. Our objective is to find a rare plant, the *sea-heath*, in one of its only Sussex localities. On the landward side of the small area of salt-marsh to the west of the Cuckmere River, look out in July and August for the small crinkly five-petalled pink flowers of this plant whose wiry stems lie prostrate on the surface of the ground. Its leaves are small and narrow and are tightly rolled under at the edges. The flowers of the sea-heath are pollinated by *hoverflies* and the seeds are probably dispersed in sea water when the tide comes high up the river bank.

Hoverflies are interesting insects and most look remarkably like wasps or bees; their appearance is an example of protective colouration or mimicry, and it prevents them from being eaten by birds which avoid bees and wasps. Interestingly, hoverflies are not about in large numbers when young birds are learning the differ-ence between good food and bad, but are common enough once the birds have learnt that black and yellow stripes mean danger. One hoverfly, *volucella bombylans*, mimics the *bumble-bee* and it totally deceives the bee into allowing her to enter its nest and lay her eggs. Normally a bumble-bee will sting to death any such intruder but the hoverfly camouflage is obviously almost perfect.

Our journey now takes us past Seaford to the important botani-cal site of Tide Mills situated just below the downs south-west of Bishopstone. Tide Mills, now a significant area of shingle beach, has an interesting history. About a thousand years ago the river Ouse emptied itself into the Channel just under Seaford Head, its mouth having been pushed by the tides and prevailing south westerly winds eastwards, a phenomenon known as longshore drift. Great banks of shingle formed between the river and the sea and parallel to the sea coast. Over the years the mouth of the river became blocked by shingle and cuts were made into the sea to prevent flooding upriver. The final cut was made so that today the river enters the sea under Castle Hill at Newhaven where its mouth once existed many thousands of years ago.

The name Tide Mills goes back to the mid-eighteenth century. Then the area around the present ruined village of Tide Mills consisted of shingle and marshland. The Duke of Newcastle, of Bishopstone Place, had the concept of building a mill for the grinding of corn into flour, operated by the action and power of the tide. The area was particularly suited to this idea of industrial enterprise on account of the geography and layout of the river Ouse and its creeks. In 1761, the Duke secured the passage of a bill through Parliament allowing three local merchants to construct the tide mills, together with buildings to house the workers. Soon a whole village consisting of Sussex boulder flint cottages, black-smiths and carpenters' shops, mill offices and a granary were constructed for about a hundred workers and their families. On the flood tide barges laden with grain entered the creeks, and water wheels drove sixteen pairs of millstones to grind the corn into flour. A windmill was also constructed on top of the granary to lift grain from the barges and lower flour down onto them, thus increasing the efficiency of the whole system.

The life of the inhabitants some 250 years ago can be imagined; no running water, sanitation by outside closets, water carried by bucket from communal taps, no rubbish collection and indoor lighting by candles. At the same time, villagers would have to contend with great storms and floods that were a permanent threat to this part of the coastline. But life went on with an average working day of 16 hours using the natural energy of two tides in every 25 hours and a weekly output of some 1,500 sacks of flour was frequently reached. However, by the middle of the nineteenth century, cheap grain from foreign countries was flooding into the country as a result of the repeal of the Corn Laws in 1846. The grain was now ground by engine-driven mills and competition was fierce. The coming of the railway to Newhaven in 1847 hastened the building of immigration and passenger facilities for the port. Newhaven expanded and in 1884 buildings were constructed over the creek so denying access for barges to Tide Mills. The mills, already severely damaged by violent storms in 1876 and 1883 never worked again after 1884.

Tide Mills continued to be occupied by workers on the railway and on the construction of Newhaven Harbour. During the 1914–18 war many of the cottages were demolished and a few years later the village was used for racing stables. The racing stable moved out in 1930 and the Chailey Heritage Beach Hospital for Crippled Children was built. As the threat of invasion grew in 1940 the children were moved out, and to prepare the defences of Newhaven Harbour, the army demolished the whole village. So by 1945 Tide Mills presented a bleak and deserted place of ruins suited only as a haunt of birds and wild flowers. Once again, as it must have been a thousand years before, this place became a haven to wildlife with the noise of ceaseless winds and the crashing of surf on the steeply shelving beach of pebbles, broken occasionally by the mechanical gnashing of a passing train on lines that extended to Seaford. I can remember as a boy in the 1950s seeing *lapwings* and *ringed plovers* nesting among the many wild plants of this bleak but aesthetically attractive place, and on the nearby wetlands, *redshank* could be heard piping their evocative alarm calls.

Now, as I write, the port of Newhaven is enlarging yet again and the whole area is threatened by developers who have already built huge warehouses upon much of the shingle shore. Many people now visit the place to walk their dogs and enjoy a certain sense of isolation, and this human disturbance has resulted in the loss of most of the breeding birds, but the wild flowers still hold on to a precarious existence.

Some of the flowers that grow in the area of Tide Mills are quite uncommon like *orange mullein, common meadow-rue, longleaf* and the *Duke of Argyll's teaplant*, while commoner plants on the shingle bank include *blue fleabane* and *rock samphire*. In places, particularly in the old east pond-bay there is a good saltmarsh habitat with plants like the *greater* and *lesser sea-spurrey* and *glasswort*.

Orange mullein, a garden escape, is somewhat similar to the indigenous *great mullein* which also grows here but has larger orange-yellow flowers. The flowers of both species bloom from June to August and are carried in clusters on tall spikes up to 120 centimetres high. Both species are covered in soft wool-like hairs and the leaves are very large and sword-shaped. Our Saxon forefathers dipped the dried stems and flowers of great mullein in melted grease and then used them as torches at mediaeval church festivals. The leaves were once used as lamp wicks and are now used in homeopathic medicine as a tincture to ease coughing. Over many centuries the whole plant has been used in many forms to ease such ailments as asthma, nervousness, neuralgia and stomach cramp. A plant with many uses; no wonder both mulleins are found around many ancient sites of human habitation.

On the banks of the wet ditches in July and August you may find the yellow flower-clusters of the *common meadow-rue*. The flowers have no petals and the bees and flies are attracted by the bright yellow pollen-bearing spreading stamens. Often more than a metre tall this striking plant has smooth, long-stalked, dull green leaves which are somewhat paler on the underside. The stems are stout and furrowed and it has a creeping yellow rootstock that sends out runners.

Longleaf is a member of the carrot family and has the same distinctive white or yellow flowers grouped together at the end of individual stems. This grouping is called an umbel and is a characteristic of this family of plants. However, what makes this particular species so recognisable is the large lance-shaped leaves which are divided into threes and are each sharply toothed. The whole plant nearly a metre tall sometimes forms large patches of somewhat straggly flowers. Look for it in wet ditches near the car park off the A259 road.

The *Duke of Argyll's teaplant* (lycium barbarum) is an attractive shrub with arching grey, sometimes spiny stems bearing lance-shaped grey-green leaves. The funnel-shaped flowers are lilac-purple, about a centimetre long and bloom from May to August. The bright red berries are relished by birds, but like the leaves and despite the 'tea' name, are poisonous to humans. This shrub is a native of China but was widely used as a hedging plant in English gardens from the eighteenth century; that is why you will see it now naturalised throughout most of England, and here at Tide Mills it can be seen all around the old village. Its intriguing name supposedly records the botanical mix-up when the labels of a true-tea plant and this lycium were inadvertently, or as a joke, transposed when the plants were given to that famous plant-collector, the third Duke of Argyll. He continued to grow the lycium with its original and incorrect label. The story did not come to light until 1838, well after the Duke had died but the plant still keeps its paradoxical title.

Walk across the large area of shingle and look for the *blue fleabane*. About 20–30 centimetres tall, this plant has an erect stem which branches out in July and August to support flower heads containing many small flowers. The outside rays of the flowers are purple with inside discs coloured yellow. The whole plant is hairy and the untoothed lance-shaped leaves clasp the stem alternately.

Another plant you will see flowering from June to September particularly around the old ruins, is the *rock samphire*. The flowers are yellow-green in colour and are formed in umbels making it a member of the carrot family, like the longleaf plant described earlier. The rock samphire, unlike any other of its family, has thick, smooth succulent leaves which are covered with a greyish-green bloom, characteristic of so many seaside plants. The succulent nature of the plant is due to the development of large cells known as 'aqueous tissue' which are employed for storing water. In dry weather, as water is lost by transpiration and by the drying effect of salt-laden winds, these water-holding cells shrink until the rains come and they expand again. Thus they act as a water storage system for use by the plants in times of drought when their immediate environment on the shingle becomes very arid indeed. The rock samphire like other maritime plants has long tap roots which burrow deeply into the ground in search of water. The nature of such plants has certainly evolved in an impressive manner to cope with the vicissitudes of life. Rock samphire though is more than just a maritime plant to be seen and then forgotten about; for many hundreds or even thousands of years this plant has been used as food. The stems, leaves and seed pods make a fine pickle if sprinkled with salt, boiled, and covered with vinegar and spices. Rock samphire leaves and roots can also be eaten with butter after having been boiled for 10–15 minutes. Shakespeare knew of the plant growing on the cliffs of Dover and mentioned it in *King Lear*, when he wrote 'halfway down hangs one that gathers samphire; dreadful trade!'

The *greater* and *lesser sea-spurrey* both grow on the banks of the salt water pond to the east of the old village. Here they flourish on the earth banks kept salty by the ingress of sea water to the pond itself. The greater sea-spurrey has white or pink flowers almost a centimetre across which bloom from May to September. Its long pointed leaves are fleshy in nature. The lesser sea-spurrey has smaller flowers only half a centimetre across and pink with a white centre. They also bloom from May to September. The lesser sea-spurrey unlike the greater sea-spurrey has small glandular hairs around the upper stem and flower-heads.

Another plant growing in the salty mud around the east pond is the *glasswort*. This plant is easily identified by its much-branched succulent stems, up to 40 centimetres high, which appear leafless but in fact the fleshy leaves are completely joined together in pairs on the stem itself. The tiny flowers appear from August to October as opposite groups of three on the leaf segment. The whole plant turns reddish in late autumn and makes a wonderful sight where it grows abundantly. Glasswort gets its name from the fact that it was formerly collected and burnt as a source of soda for soap and glass-making. The plant can also be eaten raw as a salad plant by holding the root and stripping the stem between the teeth. It can also be boiled for 10 minutes or so and eaten with butter, like asparagus.

Before leaving Tide Mills a brief word about the butterflies of the area. With such a wide variety of vegetation and habitat you are likely to see just about any of our commoner butterflies and also some of the less common. To the west of the area near the industrial estate there are many naturalised *buddleia* bushes and here in midsummer, if it is a good year for them, you will see the unmistakable *painted lady* butterfly with its colour of pale orange with black and white markings. Not many people realise that, like a number of our butterflies, this is a migrant which flies across the Channel in April and May from its main breeding grounds in North Africa and Western Europe. In some years such as 1994 and 1998 when the weather conditions were unfavourable, or when their breeding success in North Africa and Europe has been poor, very few succeed in arriving here. In other years such as 1969, 1980 and 1996 many tens of thousands succeeded in coming to Britain. With a wingspan of nearly 6 centimetres the painted lady is one of our larger butterflies. The blood and carcass of this butterfly is dark red in colour and it has been suggested that it was the remains of millions of them that turned the rivers of Egypt red as described in the Old Testament (*Exodus*, chapter 5). When resting on bare ground with its wings closed, it blends in well with its surroundings, and it craftily points its head in towards the sun so that no revealing shadow is cast to give away its position to a passing bird. The early migrants lay their eggs on a variety of wild flowers, but mainly thistles, and they hatch out in about a week into black caterpillars with yellow spines and yellow stripes down their side. These give rise to a single generation of butterflies which are seen in late summer and autumn. Unfortunately, these soon die as colder weather arrives and unlike some other species, such as the *small tortoiseshell* and *peacock*, they have not learnt to hibernate over winter.

Once when walking over the shingle waste, in early autumn with a brisk northerly wind blowing, a flash of chrome yellow suddenly caught my attention as the butterfly sped with the wind across the pebbles, then another and another. I instantly recognised a host of *clouded yellow* butterflies (described in Chapter 3) obviously intent on flying back to mainland Europe from where they originated earlier in the year. As I attempted to photograph them I noticed with some excitement that one had a much paler upper wing surface than the others. This I identified as a variety called 'helice'; some ten per cent of the female clouded yellows are of this kind. I chased this one for some time until eventually it settled on a teasel plant and I succeeded in capturing it on film.

So we leave Tide Mills and its history and people. They worked hard at a time when, without work, you were soon forgotten and neglected by those who had little time to look after others but themselves; but those inhabitants of Tide Mills did have a strong community spirit and they laboured tirelessly under the strict supervision of the owner, William Catt. They also played hard at the local public house 'The Buckle' (now a private dwelling), so hard in fact that William Catt had to impose a curfew, and every night at 10.10 p.m. precisely the three entrance gates to the village were closed and locked; so discipline was strictly enforced to ensure an efficient working environment.

Sit amongst the flowers and ruins of this village; as the hum of bees, the rustle of a breeze through the grasses and the murmur of a distant sea all combine to dull your senses, ponder awhile on what all this was like two hundred years before. What stories these ruins could tell of hardships, when ferocious winter gales brought floods and destruction, and a poor harvest meant little work at the mills. Yet at the same time there must have been much satisfaction too, when high grain prices gave job security, when the sea was calm under a great blue sky, and picnics on the beach and on nearby downs gave more than just transient joy. As those hardy people lived their lives, time, as with us, passed more quickly as the seasons came and went. Time then, as now, marked the birth of life itself, and death when it departs; before and after, time is immeasurable and without meaning. As they lived they would have been familiar with pleasure and pain, rapture and torment, happiness and unhappiness; all feelings that must inevitably be found together, for one cannot exist without the other and each is dependent one on the other. Worth thinking about as you sense and feel all around the spirits of these long-forgotten people. Suddenly a frantic shriek of a *blackbird*'s alarm call, 'tchink, tchink, tchink', awakes you from your reverie.

As you cross the main road and along the little lane to Bishop-stone a *fox* nonchalantly crosses your path and heads for Rookery Hill on your left; *rooks* still nest here as they must have done for hundreds of years. Hard to believe that in the autumn the long front lawns of nearby dwellings produce an abundance of one of our delightful wild orchids, the *autumn lady's tresses*. Even more extraordinary, only recently have we been told that beneath the hill of nearby South Heighton lay a hidden and secret military installa-tion that existed during the last war. Known as HMSO *Foreward*, Newhaven Royal Naval Headquarters, this site was actually a secret wartime underground intelligence centre. Only now with the demise of the Cold War can its existence be revealed, although local people have known of it for years but kept its secrets.

Bishopstone is a village that is mercifully preserved from devel-opment because it has no through road. It dates back at least 300 years before the Domesday Survey of 1086 and it has an equally old Saxon church. At one time the sea must have swept up the valley and perhaps the village then had its own anchorage for ships. Nearby, is a pond in which the *water-soldier* and the rampant *New Zealand pygmy weed* thrive.

The *water soldier* is a rare plant of still, lime-rich waters. It has leaves resembling those at the top of a pineapple plant, long and narrow with sharp, saw-like edges. The plant floats on the surface in summer and puts out large white flowers four centimetres across in June to August. In Britain only female flowers are known, and therefore no seeds are produced; reproduction is achieved by the growing of offshoots that break off and float away from the main body. Unusually, after flowering, the whole plant sinks to the muddy bottom to hibernate for the winter. Nothing more is seen of the water soldier until spring when it resurfaces, puts out fresh leaves and prepares for flowering again.

Another aquatic plant, *New Zealand pygmy weed*, grows in the shallow water at the edge of the pond. It has tiny four-petalled flowers held on short stalks which bloom from June to August, the plant having narrow leaves formed in pairs at intervals up the stem. The whole plant sprawls across the mud or lies submerged under the water. The plant, as its name implies, is a native of the South Island, New Zealand, and Australia, and was introduced as an aquatic plant in Essex in 1956. Since then it has spread quickly across the country from Sussex to Argyll and can now be found in many of our village ponds.

We travel on and upwards through places with tantalising names such as Poverty Bottom, Beacon Hill, Five Lords' Burgh and

Gardener's Hill. All have their own history and interesting stories. Beside these tracks across wide open cultivated fields you may come across that beautiful and now uncommon plant, *chicory*. This provided an important cash crop during the last world war when its thick fleshy tap-roots were roasted and ground and used for the making of coffee. It is still grown on the Continent for such purposes and its leaves can be used as a somewhat bitter, but tolerable addition to a mixed salad. Chicory is up to a metre tall with tough erect stems and dandelion-like leaves. Its most striking feature is the paired, large, vivid bright sky-blue flowers, which once seen, are never forgotten; there are no others like it growing wild in Britain. In places there are areas of hillside too steep for ploughing where the original downland turf still exists; here you will find many orchid species, particularly the *fragrant orchid* with its tall spikes of beautifully perfumed flowers.

The wide open skies give this area an atmosphere all of its own; a place where there are few trees, and although intensively cultivated, there are large expanses of set-aside which in May and June are full of wild flowers, such as *cut-leaved cranesbill, hound's-tongue, common mallow, wild mignonette, red clover, hop trefoil*, and *wild strawberry* to name but a few. Here also the *beaked hawk's-beard* with its yellow heads grows in profusion and in places bright red flowers of the *common poppy* make a showing as it tries to regain its dominance of forty years ago before herbicides were used so extensively. Amongst this vast array of tall grass and flowers many pairs of *skylarks* make their nests and their calls are a delight to hear on a calm evening in May when the shimmering sea gives a fitting background to their presence. Let us hope these areas of set-aside will not be cut in June or July so that the skylark and other ground nesting birds will be able to rear their young successfully. Set-aside here is clearly an important aspect of nature conservation.

We make our way by marked and unmarked tracks towards Itford Hill; always there is a distant view of cultivated fields blue with *flax* in summer and patches of scrub and rough pasture. Blackcap farmhouse, splendid in its isolation, is nearly always in sight with its tell-tale belt of tall trees, mainly *ash* and *sycamore*, a short distance in front of it. Two uncommon plants may be found along these trackways, namely *sainfoin* and *wild clary*.

Sainfoin is distinctive with its strikingly beautiful clear pink flowers veined with a deeper rosy tint. About half a metre tall it flowers from late May to August and here on our chalk-hills it is considered to be indigenous and not just an escape from cultivation. Sainfoin comes from a French word meaning wholesome hay and indeed it is grown in places as a fodder crop. Many of our motorway banks have been sown with the continental variety of this plant which has a less deep pink colour.

Wild clary flowers on dry chalk grasslands from June to September and can be recognised at a distance by its separate and apparently leafless whorls of deep purple flowers. Its erect and branched leafy stem is up to 60 centimetres tall with stalkless, toothed, somewhat rounded leaves clutching its stem. The root leaves are also toothed, are oblong in shape and have stalks, and all the leaves have a characteristic wrinkled appearance. The name 'clary' is an ancient one derived from 'clear-eye', and another old English name for it was 'see-bright'. Clearly this plant was once used medicinally and indeed friars and physicians of the Middle Ages found that the seeds of clary rapidly absorbed water and became mucilaginous. If placed in the corner of an eye, the mucilage picked up any dust particles that had blown in.

The views from Itford Hill are glorious; in May the Ouse valley below is covered in fields of yellow *buttercups*, and patches of *common reed grass* colour the side of ditches with a very pale whitish brown. This lovely tall willowy grass is a haven for the *sedge* and *reed warblers*, and a favourite nesting site for the *moorhen*.

Eastwards and past Beddingham Hill we make our way down the steep escarpment to the old coaching track between Lewes and Eastbourne that runs just south of Firle Place. As we make our descent we look away at the area of downland just east of Lewes and dominated by Mount Caburn. A *brown hare* suddenly bounds up out of the long grass and dashes uphill in a zig-zag fashion, momentarily crouching out of sight behind a *yellow hill ant*'s nest-mound before disappearing over the horizon.

Brown hares are easily distinguished from their relatives, the *rabbits*, by their larger size, browner colouration, longer ears and longer hind legs. Normally the two animals do not mix and the hares live in well-defined territories. During the day the hare lies well camouflaged under cover in shallow depressions known as 'forms'. In March the males stand on their hind legs to box each other in a ritual that impresses the females before mating; this habit has given rise to the expression 'the mad March hares'. The females give birth to two to four young (known as leverets) up to four times per year. They are born above ground with a full coat of fur. The female, or doe, as she is called, sometimes distributes her young in several separate forms which she visits at intervals to feed them her milk. Adult hares feed mainly at twilight on grass roots and the produce from cultivated fields and gardens. Brown hares have become scarcer on their downland habitat over recent years, possibly because there is less natural ungrazed grassland for them in which to survive. Another reason could be that they are now more often shot for their meat as they are considered to be more of a delicacy, especially since myxomatosis has given a certain distastefulness to the rabbit. The hare of course is not affected by that disease.

The old coaching way is well defined by a rutted chalky track that is marked on the modern Landranger Ordnance Survey map as a 'byway open to all traffic' but little except mountain bicycles can use the deeply rutted parts of it now. Beside this ancient track grow huge beech trees some 200 years old whose canopy in summer gives shade to those who pass along the way. To the east of Firle Place along this track can be found the remains of a group of old dwellings, once called Beanstalk Cottages, together with a public house called The Beanstalk. The latter is now a private house and all that remains of the former are some old walls and foundations. This place and its associated stables were once used as a resting place for the coachmen and horses of the last century on their way to Eastbourne and Pevensey. An interesting plant still grows around the ruins; *elecampane*. This plant is not a native of Britain but is an Asian species introduced into our country by the Romans for its medicinal properties. Elecampane is a tall plant, growing up to a metre and a half high. In July and August its tough stems bear heads of yellow flowers some 8 centimetres across looking something like small sunflowers. The broad, lance-shaped leaves of elecampane are large and saw-toothed and are covered in a soft white velvet underneath. These leaves have a soothing effect on limbs and were used to cover the lower legs of the coaching horses after a long hard ride. For this reason the plant is often found near places associated with horses. During their occupation the Romans found Britain cold and damp and used to dry the roots of elecampane, covered them in honey or sugar, and ate them as a treatment for asthma and other breathing difficulties. More recently the chemical inulin has been found in this species as well as sunflowers and this is now used in the treatment of asthma.

As you rest among the ruins of Beanstalk Cottages think of those times long ago when this was a busy place; coachmen seeing their passengers into the warmth and light of the tavern, while liverymen fed and stabled the horses for the night using ample supplies of elecampane leaves to soothe those that were lame and tired. On a cold wet night in winter the sighs and sounds of those people who worked so hard in this place are easy to imagine in the moan of the wind as it tears at the huge *beech* trees nearby.

Mount Caburn beckons us on the far side of the busy A27, so make your way to Glynde for the easy climb to the top of Caburn via a path that starts opposite the entrance to Glynde Place. Just before starting the climb, walk a short way along Ranscombe Lane leading to the farm of the same name just below the old Roman encampment. Just south of the lane near Brigdens Farm is a field in which grows a rare plant, the *small-flowered buttercup*. This species is rapidly decreasing as more and more of its habitat of dry sandy exposed ground is destroyed and herbicides are sprayed on its other favoured location of grassy banks beside roadways. Flowering in April and May it can easily be recognised by its pale yellowish-green leaves, its furrowed and hairy stalks and its small yellow flowers whose petals are dwarfed by the bristle-covered immature fruits. The whole plant spreads over the sandy and grassy bank here.

The path up to Mount Caburn is an ancient one, possibly used by the womenfolk of the Iron Age camp as they collected water from nearby springs in earthenware jugs. Part of the trackway where it is deeply sunken is now overgrown with hawthorn and bramble and is a favoured nesting site for the *whitethroat*. In May the vigorous, urgent chatter of these birds and their repeated 'check, check, check' is nearly always apparent. At the same time of the year you will see the two little known butterflies, the *dingy skipper* and the *grizzled skipper*, darting with fast wing-beats amongst the grasses.

The *dingy skipper* is so named because of its dull, brown colour with a grey fringe around its hind-wings. When resting in sunshine during the day it basks with its wings spread out flat, quite unlike other skippers which rest with their partly raised wings ready for a quick take-off. The *grizzled skipper*, described in the previous chapter, is more conspicuous, with its easy-to-identify markings of dark brown and white speckles with evenly spaced dark and white patches around the edges of the wings. The dingy skipper's eggs are laid singly in early summer on the food plant of the caterpillar, *bird's-foot trefoil*. The caterpillar overwinters as a caterpillar inside a cocoon it has spun inside a tent of leaves. It turns into a chrysalis in April and the butterfly emerges about a month later in May. The grizzled skipper overwinters as a chrysalis held in a cocoon of silk and leaves and this hatches out in May and June. Both species usually produce only one generation a year and each butterfly only lives for about two weeks. Watch them as you slowly climb the hill and see how they use a tall grass stem as a look-out post. They both aggressively defend their territory and will suddenly fly off to intercept intruders such as other skippers, or flies and bees.

Later in the year, this track is alive with blue butterflies of many kinds, *marbled whites* and *red admirals* to name but a few. Wild flowers of many species grow abundantly all around and one that stands out well is *crosswort*. About 40 centimetres tall, this is very distinct from all other species by reason of its large egg-shaped and pointed hairy leaves arranged cross-wise in a whorl at intervals up the stem. The four-petalled small yellow flowers are formed in clusters just above the leaf whorls and appear in April and May. The flowers have a strong smell of honey and are attractive to pollinating insects.

As the summer months approach many more flowering plants appear. The *tall melilot*, up to a metre and a half high, puts out its many long-stalked sprays of pea-like, deep yellow flowers from late June to September. Its leaves are broken up into three slender oval leaflets with toothed edges. Tall melilot is one of three similar species that were introduced from Europe as medicinal herbs and fodder plants, being used as poultices and to make 'melilot' plasters.

A huge area of set-aside on both sides of the path greets you as you reach the ridgeway and before you turn left for the summit of Mount Caburn itself. This set-aside area is marvellous in its simplicity, an area allowing for a natural regeneration of plants without interference by man. In May it is a blaze of yellow of the *beaked hawk's-beard* intermingled with *common poppy*, *thistles*, *vetches*, and many others. Here is a haven for the *skylark* and *meadow pipit* and both can be seen flying out of the tall vegetation, busy nest building, or even feeding an early brood of young. A flock of *rooks* and a solitary *carrion crow* can be seen marauding through the fields intent on finding anything they can eat from a small insect to a nest of eggs or young. A month or so later the set-aside provides cover and a sanctuary for many butterflies such as the *common blue*, and the *chalk-hill blue*.

The approach to Mount Caburn is obvious, a white streak across and over the defensive north ditch indicates the path into the ancient site. A sign indicates that you are entering a nature reserve, and what a contrast, for as you pass through the swing gate a herd of inquisitive bullocks greet you. The grass is short, trampled and covered with the excrement of cattle; no haven here for flora, birds or butterflies. Over the Iron Age encampment itself and just below are over fifty sheep grazing the southern slope. The grass, because of this grazing, is short and lacks the luxuriance and vitality of the set-aside. Certainly I remember it 20 years ago when there was much less grazing with fewer sheep, and the vegetation was 30–60 centimetres high, butterflies of all kinds abounded and skylarks were all about. Who has allowed this once-wonderful area of chalk grassland, awash with tall flowers, to become but a shadow of its former self; for now it is difficult to find much that was once common on its slopes, such as the *burnt orchid* and the *chalk-hill blue butterfly*?

The *burnt* or *dwarf orchid* is only 6–10 centimetres high and is not easy to find. In May and June its flowers are borne on a single stem and have a dark purple hood and a lip which is white and speckled with red. The unopened flowers at the top of the spike are a rich red-brown and give this orchid its name 'burnt'. Once quite common on the lower slopes of Mount Caburn it is now more common on the slopes a kilometre or so to the north-west, towards the golf course.

Once, the *chalk-hill blue* butterfly was seen in large numbers in July on the lower and more sheltered slopes. For some years now, they, like many others, have been absent. The reason for this I am sure is the destruction of their habitat. They like tall chalk grass-land rich with flowers like *trefoils*, *vetches* and *marjoram* and are unhappy on very short grass devoid of shelter and cover. Sadly short grass predominates on the southern slope of this hill in midsummer. Even as I write in June after a very wet spring the slopes are still being grazed by sheep and the grass has no chance to grow into the habitat so beloved of many of our downland butterflies. Of course I am conscious of the legacy of John Ellman of Glynde and his famous breed of Southdown sheep, but if we want to preserve our natural grasslands we must compensate farmers for not grazing their livestock on sensitive areas between March and October.

One of the more attractive shrubs to be found on Mount Caburn is the *burnet rose*. This somewhat local plant grows some two metres tall and has creamy white flowers sometimes tinged with pink which have a very sweet smell of honey and jasmine. It flowers from May to July; look for it on the lower part of the south-facing slope. Surprisingly, some of these shrubs have been deliberately taken out and destroyed recently, possibly to reduce the cover for *rabbits* which are also causing habitat destruction. Nearby, in the hedgerows and beside the path that leads down to Glynde, you may find another but very rare rose, the *small-leaved sweet briar*. Its white to pale pink flowers bloom at the same time as the burnet rose, but it can easily be distinguished from that species by having fewer thorns, and by the fact that if its less rounded leaves are crushed they put out a characteristic apple smell.

On top of Mount Caburn grow some interesting plants that to some extent can tolerate short close-cropped grass. The *field fleawort* described in the last chapter and associated with Iron Age sites can be found in May on the western edge of the site itself. It is affected, however, by the rough treatment it receives from the numbers of hang-gliders using the site from time to time. I am very aware of the pleasure this sport gives to people, but is it really sensible to allow these aerial enthusiasts to pursue their interests on a nature reserve? The two are hardly compatible and I would have thought there are many other less sensitive areas nearby that could be used, Beddingham Hill and Itford Hill for example.

At the bottom of the defensive ditch look for *mugwort* and *silverweed*. Mugwort grows up to a metre and a half tall with deeply cut leaves which are greyish below and glossy green above, its tiny reddish-brown flowers appearing in July and August. Mugwort is a plant that is common along our roadside edges and in hedge bottoms and it is something of a mystery why it grows so commonly on the top of Mount Caburn. Maybe it was deliberately introduced here by the Iron Age Celts since the plant has long been associated with magical properties and was believed to be a protection against thunder and witchcraft. It was also said to offset fatigue and travellers once stuffed it into their shoes to prevent travel weariness. In the past it was sometimes used as a leaf smoked by children.

The leaves of *silverweed* are easily recognised with their silky green above and silvery white underneath. The plant grows in dense patches closely hugging the ground and is particularly abundant where moisture can collect in depressions and hollows as here in the bottom of the defensive ditch. The solitary five-petalled yellow flowers are some 2 centimetres across borne at the end of long stems and appear from June to August. The roots of silverweed are edible and the plant was once cultivated in the west of Scotland before the introduction of the potato; the roots being boiled and dried and then ground into flour for bread or porridge. Like mugwort, silverweed was often stuffed into boots to ease travellers' aching feet. Silverweed's Latin name, *anserina*, from *anser*, a goose, refers to the fact that geese are very fond of eating it and that is why it is commonly found around ponds.

Many other plants grow on and around Mount Caburn, so see

how many species you can find and identify when you walk upon its soft turf. Watch the *fox* stalking *rabbits* on its lower slopes and see the *kestrel* hover in the wind as it seeks out its prey of *vole* or *fieldmouse*. On a still day in spring, the excited chattering of *jackdaws* echoes from the sides of the nearby chalkpit where the birds are building their nests. As you sit on top of this beautifully shaped hill let your mind wander back to the time when the sea lapped round its southern extremities, when ancient man watched porpoises playing in the sea below. The unrelenting and voiceless wind sweeps over the top as you take in the wide panoramic scene all around and you consider your next move over the hills to the north, past the golf course and down into the nature reserve of Malling Down.

On your way you walk over good areas of chalk grassland with a significant variety of wild flowers. Interestingly in some years the *burnt orchid* grows in large numbers on these slopes flowering in May. The *corn bunting* is often seen in this area sitting on a fence post singing its distinctive song, a rapid dry jingle like the rattling of a bunch of keys.

Malling Down Nature Reserve is managed by the Sussex Wildlife Trust and consists of a series of ancient pits where chalk was once quarried, and a deep combe valley, the top of which is dominated by Malling Hill, another area of rich grassland. Among the many downland flowers of the short grass around the chalk-pits is the *musk orchid*. Despite its name, this orchid is sweetly scented and does not smell of musk. The scent is particularly strong in the evening after rain and attracts many small insects such as flies and beetles which carry the pollinia to other flowers to effect pollination. Each plant has a main tuber from which the flowering stem arises, and two or three others at the end of horizontal fibrous roots. Each year more tubers are produced and by this means it is able to reproduce itself vegetatively as well as by seed and is also the reason why it is seen growing in clumps. Each flowering spike is only 5–15 centimetres high and is densely packed with 20 to 30 tiny greenish-yellow flowers which appear in late June and July. Because of its colouring and diminutive nature it is easily overlooked. As with many other orchid species, in some years on Malling Down there may be tens of thousands of flowering spikes and in others there may be just a few.

In July and August, the *carline thistle* makes an attractive addition to the chalky mounds. It has a beauty of its own with its yellow-brown flower heads which are each surrounded by yellow bracts (not petals) which appear like rays. These close up over the flower heads in damp weather in a protective cover. Carline thistle has spiny, deeply lobed leaves and grows 5–30 cm tall. The whole plant has an everlasting and somewhat striking appearance and in winter, long after the flowering period is over, you will see it standing out amongst the dead grasses and vegetation that surround it.

Summer at Malling Down brings with it many species of butterfly, the most important of which is the *Adonis blue*. The male Adonis blue is the brightest and most vivid blue of all British butterflies, and is, quite aptly named after Adonis, the god of masculine beauty. Once seen here on the warm south-facing slopes where the grass is short it is never forgotten. The female is not so adorned as the male and is brown in colour with just a hint of blue where the wings come out from the body. Both sexes have a chequered white band around all the wings. The food-plant of the Adonis caterpillar is the horseshoe vetch under the leaves of which the eggs are laid singly in May and again in August. The eggs hatch out after about 30 days and the caterpillars, which have green and yellow stripes, feed at night on the host plant. The summer caterpillars turn into chrysalids after about two months, and some three weeks later these turn into adult butterflies. The later caterpillars hibernate in winter, form chrysalids in April, and develop into butterflies in May. So there are two generations of the Adonis butterfly, one flying in May and June, and the other in August and September.

Like some other blue butterfly species the Adonis has a strong symbiotic relationship with the little *yellow hill ant*. The winter hibernating caterpillars are taken by the ants into their nest where they are protected from predators; in return they produce a sweet secretion on which the ant feeds.

The Adonis blue butterfly requires certain types of plant such as *vetches* and *marjoram* to feed on, as well as a specific type of habitat, i.e. a chalk grassland slope facing south or south-west. This combination is hard to find, and consequently the Adonis is sadly a declining species.

Another butterfly I have seen recently on Malling Down in August is the rare *silver-spotted skipper*. This butterfly has a fast flight and keeps close to the ground. At rest it can be distinguished from other skippers by the silver patches on the greeny-brown underside of the wings. Like the Adonis this species requires a certain type of short chalk grassland with a southerly aspect giving plenty of sunshine. The caterpillar of this skipper has evolved an unusual and fascinating habit. It feeds on grasses and when it feels the hot breath of a grazing animal it immediately curls up and drops to the ground, where it escapes the danger of being eaten. Let us hope that my sighting of this butterfly was not just a random one, but that it is now firmly established on the reserve as a breeding species.

Aerial photographs of Malling Down in the 1950s show little hawthorn scrub and few trees around the combe valley and chalk pits. Grazing decreased at this time and scrub then began to grow over much of the slopes. The Sussex Wildlife Trust has removed much of the scrub, and by reintroducing sensible grazing techniques it has managed to preserve this wonderful and incomparable habitat for so many important plant and insect species.

At night in July walk over the grassy mounds of Malling Down and you may see pinpricks of pale yellow-green light appearing all over the ground. This light comes from the luminous organs situated under the tail segments of the female *glow-worm*. To attract the male, she climbs up onto a low plant and turns onto her side so that her underparts are directed sideways. The eyes of the male glow-worm are very large and are adapted so that it can easily see the glow given out by the female. The male can fly like a beetle, but the female has no wings and is more like an overgrown beetle larva. During the day both sexes hide under stones and rubbish, do not feed, and live for only about nine days. The larva of the glow-worm feeds on slugs and snails by injecting them with a digestive juice which enables it to suck up all the fluid remains.

A tree that is common on chalky soils but is less well known than many of our other native trees is the *whitebeam*. The name is derived from the Saxon word *beam* meaning tree, and the white from the conspicuous white underside of the leaves. This whiteness is caused by felted hairs that prevent water loss from its leaves on the dry chalk soils; the leaves vary in appearance but generally are oval-shaped and toothed and set singly on stalks. These leaves first form in April as goblet-shaped groups of white, resembling blossoms, and as they open the whiteness is hidden below the bright green of the leaves' upper surface. The white flowers of the whitebeam, some 5 millimetres across, appear in May in loose flattened clusters; to be followed in October by yellow brown berries which the birds eat, so scattering the small, hard brown seeds. The whitebeam lives for about 80 years but seldom exceeds 18 metres in height. The wood of this tree is hard and of small grain and is particularly useful for joinery and small cabinet work. In a brisk breeze the whitebeam presents a splendid sight, especially at twilight, almost as if thousands of white candles were blowing in the wind. In the autumn fall, the pale grey leaves strew the floor below the tree creating almost a mystic, purplish-grey sheen. See if you can spot the whitebeam on the Malling Down Reserve.

As summer turns to autumn, and bales of straw litter the golden yellow fields below, so new wildflowers appear on the slopes of Malling Down. One particularly catches your eye in a haze of deep blue, the *devil's-bit scabious*. About 30–80 cm tall, this plant has roundish, violet-blue flower heads and narrowly elliptical, sometimes slightly toothed, leaves. It blooms in large patches from August to October. The roots of devil's bit scabious were once boiled in wine and used to treat snake-bite and many diseases including the plague. The roots are short and come to an abrupt end, giving rise to the saying by ancient friars and monks that 'the root was longer until the devil bit away the rest from spite, envying its usefulness to mankind'.

From the top of Malling Down, look across at Lewes Castle with all its history and untold stories of the past. The river Ouse winds its way down from the Weald, its gently flowing waters passing through meadows full of flowers creating places of great peace and tranquillity. Here and there these placid waters reflect the shade of *alder* trees which intermingles with the bright yellow flowers and rounded flat leaves of *water lilies*. With this peace a sense of timelessness can be found too; a scene captured so well by Delius in the second movement of his *Florida Suite* called 'By The River'.

To the west, as the small puffs of cloud turn a rosy pink with the lowering sun, the escarpment of the South Downs stands out unyielding against the evening sky and reminds you of how much more there is yet to explore.

Firle Beacon

Rodmell Church and Kingston Hill

Blackcap

83

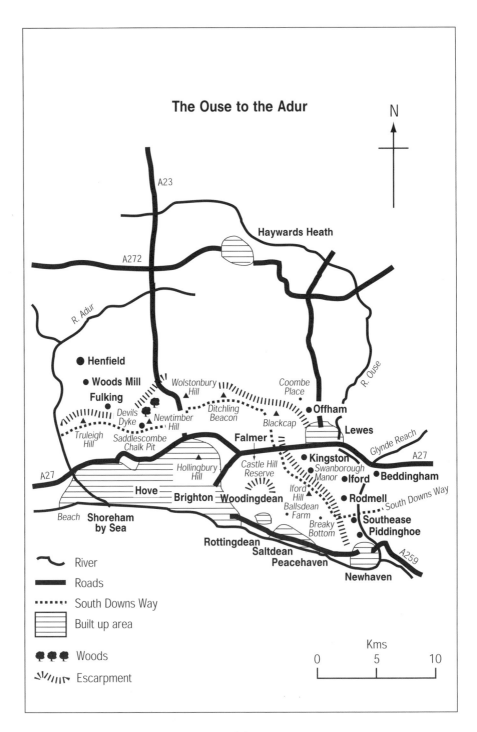

The Ouse to the Adur

N

A23

A272

Haywards Heath

R. Adur

R. Ouse

● Henfield

● Woods Mill

Fulking

Wolstonbury Hill

Coombe Place

● Offham

Devils Dyke

Newtimber Hill

Ditchling Beacon

Blackcap

Lewes

Truleigh Hill

Saddlescombe Chalk Pit

Falmer

Glynde Reach

A27

Hollingbury Hill

Castle Hill Reserve

● Kingston

Swanborough Manor

● Iford

● Beddingham

A27

Hove

Brighton

Woodingdean

Iford Hill

Ballsdean Farm

● Rodmell

South Downs Way

Beach

Shoreham by Sea

Breaky Bottom

● Southease

Piddinghoe

Rottingdean

Saltdean

Peacehaven

Newhaven

A259

River

Roads

South Downs Way

Built up area

Woods

Escarpment

Kms

0   5   10

84

5

THE OUSE TO THE ADUR

Ditchling Beacon

Mount Harry, Blackcap, and Ditchling Beacon stand proud as high points on the great north escarpment of the South Downs between Lewes and the A23 road from Brighton to London. Amidst ever increasing housing development to the north and south, this stretch of downland retains dignity and grace and offers places where one can still find solitude and peace. Our exploration of this area starts at Offham, climbing over the north side of the chalk pit by way of a public footpath. A great beech wood clings to the side of the escarpment above Coombe Place. In the 1950s a footpath passed through the wood with easy access from the B2116 at Courthouse Farm and beside it one could find *fly orchids* and *bird's-nest orchids*. Now the access has gone, the wood is private, and only a few fly orchids exist precariously, in places where brambles have not overwhelmed and subdued the delicate balance of the beechwood flora. The fly orchid is a plant that thrives best on the edge of woodlands with bare patches of fallen leaves and some mossy growth and although it can tolerate heavy shade and can grow through low cover of other plants such as *dog's mercury*, it cannot survive in thick undergrowth or beneath heavy leaf canopy.

The *fly orchid* has a stem 15–60 centimetres in height, which bears two to ten somewhat small flowers. The upper petals which resemble a pair of antennae are purple-brown in colour and slightly velvety. The dark red purple labellum is long and narrow with three lobes, and has a band of brilliant iridescent blue across its centre. Just below the upper petals are two dark glistening patches of nectar looking like eyes, and it is these that attract the insects for pollination. When the flowers have been open for a day or so they turn dark brown, then yellow-brown and the central band becomes a dull white. Flowering in May to early July, the fly orchid gives us the best impression of an insect resting on a plant, and as we shall see in later chapters, this orchid like many others is prone to fertilisation by other species to form what is called a hybrid plant.

The whole line of hills stretch out to the west and as you walk the ridgeline along the South Downs Way, often you may see nobody and seem utterly alone. However, you will never be lonely, for here you will have the company of many birds including the *corn bunting* and you may catch a glimpse of a *stoat* bounding across the edge of a field.

You will most probably first catch sight of the *corn bunting* perched conspicuously on a fence post bordering the cultivated fields at the top of the escarpment. It is a heavily built bird with dull colouring of cream buff and sober browns. The corn bunting is associated with arable farming and it feeds on seeds, caterpillars, ants, and just about any vegetable or animal matter it can find. In winter, corn buntings are often seen in small flocks mingling with other birds much as *sparrows* and *yellowhammers* as they search for food. At one time, forty years ago, it was a relatively common species. However, with the increased use of insecticides and herbicides up to the mid-1980s, and the change to autumn ploughing with the consequent loss of weed-strewn fields of winter stubble, the corn bunting declined in numbers quite alarmingly. As controls were put on the use of harmful chemicals on our fields, so the corn bunting population stabilised. As I write, the implementation of set-aside land has become more widespread, and the species is now on the increase.

The male corn bunting has an extraordinary polygamous habit, and may have up to seven hen birds as mates. No wonder he takes little part in nest building, incubation or feeding young, as he is always kept busy defending his territory against other males. So it is the female who builds the nest of grass stems, lined with some hair and finer grass. The nest is placed about a metre off the ground in the base of a shrub or hedgerow. Three to five pale buff eggs, blotched and streaked with dark brown, are laid in June or July, which is late compared to most of our breeding birds. The young hatch out after only about 12 days and they are fully fledged and fly out some 10 days later. Two broods are normally produced by the corn bunting but cases of three broods have been recorded.

The *stoat*, with the possible exception of its smaller relative, the *weasel*, is the most common of our carnivorous animals. However, they are so quick and fleet of foot that they are not often seen at close quarters. Stoats have slender bodies about 32 centimetres long and, unlike weasels, have black tips to their tails. In summer, their coats are red-brown on top and white underneath, and they moult in the spring and autumn. In Northern England and Scotland, their coats turn completely white after the autumn moult, but in the south they produce a winter coat which is similar to that in summer but is thicker.

Stoats are often killed by farmers and gamekeepers because of the damage they do to poultry and game, but with more intensive farming methods of recent years they are tolerated for the good they do in keeping down vermin such as rats, voles and mice. Stoats are also welcomed as a means of reducing the *rabbit* population, which is now largely immune to the myxomatosis of the 1950s, and has hugely increased in recent years as a result. It will also eat birds and reptiles and with its excellent hearing and sense of smell it will pursue its prey relentlessly, even climbing trees and swimming across rivers and lakes. Once in direct contact with its prey, the stoat applies a powerful bite to the back of the neck, killing it instantly. They make their dens in hollow trees, crevices and old rabbit burrows, always in or near wooded areas. The females produce about six young in the spring which remain with their mother for some time after weaning. The young hunt in family groups and can sometimes be seen at play on the edge of downland woods, chasing and boxing each other, and turning somersaults.

The *weasel*, smaller then the stoat but with similar colouring, is about 20 centimetres long with a 5 centimetre tail. It hunts chiefly at night and preys on *voles*, *frogs*, *rats* and *mice*, and will also climb bushes in search of birds' eggs and nestlings. It is small enough to get into the nest boxes of tits and will devour all the contents. Occasionally it will kill larger prey such as *rabbits*. The weasel makes its nest of grass or leaves in a shallow hole, tree stump or haystack, and two litters of about six young are produced each year. You will often see weasels at night bounding rapidly across a country lane caught in the headlights of your car.

The main enemy of the stoat and weasel is man, but both animals are preyed upon by owls and hawks.

The South Downs sweep westward with many interesting areas of short downland turf where various wild orchids such as the *twayblade, common spotted* and *fragrant* grow in the late spring and early summer. Plants such as *squinancywort* and *harebell* are common here, particularly on the old chalk excavations that form little dells and mounds on the slopes above Westmeston.

The *twayblade* is a most adaptable orchid and flourishes in a wide variety of habitats, from moist woodlands to hill pastures and dune slacks. On the South Downs it is frequently found on the chalk grasslands be they of short grass or long. It has a stout stem 20–60 centimetres high, densely covered with glandular hairs, which supports a spray of about twenty small yellow-green flowers which bloom from April to June. As the name suggests it has two large broadly elliptical leaves opposite each other 6–10 centimetres from the base of the stem. Look closely at the flowers and you will see that the lip or labellum divides into two lobes. At the base of this labellum you will notice a groove down which nectar is secreted. Pollination is effected by small flying insects which visit to lick up the nectar from the groove. In crawling up the groove the insect's head is instantly smeared with a sticky glue and pollen. It takes fright and flies off to another flower, fully laden with pollen in the right place for effective fertilisation. A high proportion of the twayblade flowers produce ripe seed capsules, but the seeds take four years to develop into the characteristic two-leafed plant and it is a further ten years before it flowers. So you will often see many plants with just two leaves and no stem. This orchid can also reproduce itself vegetatively by developing buds on its fleshy, fibrous roots, which shoot off in all directions around the parent plant. This method of reproduction often takes place in woods where it is too dark for flowers to bloom and where again you will often find plants with just two leaves.

The *common spotted-orchid* is another very adaptable species and is equally at home on the chalk grasslands of the Downs as well as along roadside verges and damp meadows. It has a straight, slender stem, some 15–45 centimetres high, which supports a cylindrical spike of densely packed flowers which open in May or June and last for a month or so. The flowers are pale-lilac in colour, sometimes almost white, and the three-lobed labellum is marked by a double loop of dark purple lines and dots. These lines and dots vary greatly in intensity: they may only be slightly deeper than their background colour, or they may be dark purple, almost black. The size and shape of the labellum lobes are important in distinguishing the common spotted orchid from the *heath spotted-orchid*, which is rare on calcareous soils, and grows more commonly on moors and acid heathlands. The labellum of the heath spotted-orchid is very broad and flat, with the central lobe being small, triangular and often shorter than the lobes either side. The labellum of the common-spotted orchid, on the other hand, has a long, narrow central lobe, separated from the side-lobes by a deep cleft on each side. There are as many as 20 lance-shaped leaves at the base of the common spotted-orchid stem and these are usually spotted and blotched with deep purple, although occasionally plants with unspotted leaves can be found. The heath spotted-orchid has similar leaves but are generally more narrow, never being short or rounded, as with the common spotted species.

As with so many orchids, the common spotted frequently hybridises with others of the same family such as the *frog, fragrant*, and *marsh* orchids. These hybrids make proper identification very difficult and confusing, and most of the field guides do not make a distinction between the many sub-species.

The *fragrant orchid* is easy to distinguish on account of its rosy-pink coloured flowers which are strongly sweet-scented. The flowers, which do not have any markings on them, are held fairly densely in a spike on a 15–30 centimetre tall stem. The three to five long, narrow basal leaves are unspotted, while there are another two or three similar leaves which clasp the stem. The flowers appear in June and last until the end of July, so look for it amongst the grasses of the downland slopes, particularly where the common spotted-orchid grows.

A common plant found on the dry grasslands of the Downs and particularly on ant-hills is *squinancywort*. This is a small prostrate plant with very narrow leaves, arranged in fours around the slender stem. The small funnel-shaped flowers which bloom in June and July are white, and pinkish on the outside. The name is devised from the word 'quinsy' and was once used as a gargle for sore throats.

The *harebell* stands out well in the grey-green grasses of summer with its beautiful bell-shaped flowers of sky-blue. These graceful nodding flowers are held on slender stems and appear in June and bloom on until late in October. The stem leaves are long and narrow while those at the base of the plant are heart-shaped. Look for it beside trackways and on the short turf of the old chalk pits which were formed over the centuries by landowners and tenant farmers. These people dug out the chalk to spread on areas particularly on the top of the downs, where acid soils from the last glacial period overlie the chalk. Raw chalk was spread over these acidic soils to sweeten and mellow it for corn production and this chalk took over a year to dissolve and effect its purpose. Chalk pits were dug in many places all over the Downs and these still survive as large indentations on the escarpment, or as small circular hollows in cultivated fields on the top; these age-old chalk pits make particularly good sites for chalk-loving plants.

At Ditchling there is a magnificent nature reserve, owned by the Sussex Wildlife Trust, which encloses many ancient chalk pits and also includes Ditchling Beacon, which at 248 metres is the third highest point in East and West Sussex. Early Iron Age earthworks, in a roughly rectangular shape, can be found here at the summit, and deep trackways carve their way down and across the escarpment. These sunken ways, or bostals as they are called, are of great antiquity, being used over the centuries as cattle ways, military roads, coaching tracks and footpaths. Such bostals serpentine up the slopes over many of the Downs to disappear tantalisingly over ridgeline crests.

Two of the many interesting plants to be found on or near the reserve are the *spiny restharrow* and the *rosebay willowherb*.

Growing up to 60 centimetres tall the *spiny restharrow*, a member of the pea family, has erect spiny stems, from which grow pink flowers from June to September. The spiny restharrow thrives on rough grassland on the heavy clay soils which accumulate in places particularly at the top of the Downs. Its close relative, the *common restharrow*, which is similar in appearance but does not have the spines, is also found on the Downs, particularly near the sea. It is a more recumbent plant and has a tough underground root system which used to delay the passage of horse-drawn ploughs or harrows; hence its name. When the leaves of either restharrow are eaten by cows, the resultant milk becomes tainted with an unpleasant goat-like smell and this is particularly pronounced when turned into cheese, and such cheese is called 'cammocky'. This smell is most apparent when the leaves are rubbed and crushed together between the fingers. In Sussex and Hampshire, one of the local names for restharrow is 'cammock'. In ancient times the young leaves of restharrow were preserved in vinegar and used as a sauce to be eaten with meat, and the roots were also dug up and chewed as liquorice. A liquid extract of its flowers and

90

roots were once used by herbalists to treat disorders of the urinary tract. In July and August a blaze of deep rose-purple catches your eye just to the west of the beacon summit. This is *rosebay willow-herb*, also known as *fire-weed*, because it sometimes appears in great profusion where fire has destroyed the underlying vegetation the previous year. Growing to a metre and a half tall, the rosebay willowherb has smooth unbranched stems, each supporting a spike of elegantly spaced rosy flowers which are nearly 3 centimetres across. After fertilisation, egg-shaped seeds form, each terminating in a tuft of most delicate, long silken filaments which buoy them up as the wind scatters them far and wide. This beautiful and striking plant forms dense patches and is often found in woodland glades which have been cleared and where light is able to penetrate the canopy. However, here on Ditchling Beacon, it flourishes on the bare hillside, well away from the scrub and woodland.

Many other wild flowers are to be found on the reserve such as the *greater butterfly-orchid*, *lady's bedstraw*, *round-headed rampion*, *yellow-wort* and *small scabious*, to name but a few. Apart from the rich variety of downland plants this nature reserve is also well known for its insects. In August and September look out particularly for the *great green bush cricket* and *dark bush cricket* both of which are very distinctive.

The *great green bush cricket*, the largest of the species in Britain, may be found just about anywhere on patches of bramble or gorse, or amongst the summer grasses. Just over 4 centimetres long, commonly, but mistakenly called the great green grasshopper, it is in fact a bush cricket with its long wings and antennae, and even longer powerful hind legs. Bush crickets eat both plants and soft-bodied insects and in turn they are preyed upon by birds and rodents; but their green colour which gives them concealment in the thick vegetation, and their secretive habits, both help them to survive. Only the male of the species produces the characteristic strong, shrill, 'chirrup', as it scrapes its left forewing across the edge of the right forewing. This burst of song helps the male to attract females and warns off rivals. Autumn sees the eggs laid singly on rotting wood or in bark crevices, and these hatch out into young bush crickets in April without going through a chrysalis stage.

The commonest of our crickets is the *dark bush cricket* and on summer evenings the surrounding vegetation becomes alive with their chirring sound. Their very long antennae, their dark brown or even black colour and body length of 2 centimetres distinguishes them from other species. The female is particularly recognisable with its scimitar-shaped egg-laying organ which it uses to prise open cracks in rotten logs or under bark.

Snails are common on chalk downland slopes as the calcium rich soils provide plenty of raw material for them, with which to construct their shells. They belong to the group of animals known as molluscs, which also includes octopuses, squids, oysters, and limpets. All are descended from the same ancestor, a mollusc, which once crawled over rocks in the shallow water of ancient seas. There are 120 species of snails in the British Isles, eight of which live on land and the rest in fresh water.

The characteristic of all snails is the whorled shell that they carry on their backs which is their home into which they retreat when danger threatens. The body of the snail is a bag-like structure coiled up within the shell; only the head and bottom part, known as the foot, are visible. The laborious movement is the result of the passage of muscular waves along the foot, which secretes a trail of slime as it goes. The head is merely a continuation of this foot, distinguishable by two pairs of tentacles, the longer of which have eyes at their tips. Freshwater snails have only one pair of tentacles with eyes at their base. Being slow-moving, snails generally feed on plants, fruits and decaying matter. They have a large area in their intestine where a huge amount of food can be ingested, and this can then be digested at leisure from the safety of their hideaway under stones, logs or leaf litter. They can go for a long time without food, but they do need water to maintain the moisture in their bodies. For this reason they are nocturnal animals, emerging when the nights are damp with dew or rain. In dry weather they retreat into their shells and plug the aperture. Snails can live up to ten years but most die within two years, notably after they have spawned eggs. They are hermaphrodite (bisexual), which is a distinct advantage to these slow-moving and lethargic animals as it increases their chances of meeting a suitable partner. Courtship consists of the two animals pressing their bodies together and rocking to and fro with their tentacles touching. Mating occurs when each snail pushes a harpoon-like dart, half a centimetre long, into the foot of its partner. These 'love darts' trigger off the exchange of sperm. After mating, the snails part and each lays white, almost translucent eggs in batches under stones or in a shallow depression in the earth.

Two snails you may find around Ditchling Beacon are the *white-lipped banded snail* and the *garden snail*. The colourful white-lipped banded snail is about 15 millimetres tall and has a white, pink or yellow shell often boldly banded with dark brown. It is quite common in rough vegetation and it is said that sheep are very fond of eating this snail, and this gives Southdown mutton its special flavour! The garden snail is much larger, some 2 centimetres high, and is cream-coloured with brown streaks; it is common all over the Downs.

Ditchling Beacon Nature Reserve with its rich variety of plant life is noted for the number of butterfly species that inhabit it; two that are worthy of mention here are the *green hairstreak* and the *chalkhill blue*.

May and June are the best months to see the *green hairstreak*, so look out for a flash of metallic green around the scrubland surrounding the chalk pits. This butterfly is seldom seen with its wings wide open and it is the underside of the wings that is coloured a bright green giving it perfect camouflage when it is at rest. The upper sides of its wings are brown in colour. There are four other species of hairstreak in Britain and they are so called because of the hair-like line across the underside of the wings, but on the green hairstreak this is reduced to a row of white dots. The green hairstreak is the only British hairstreak that does not have small 'tails' on the hind-wings. Green hairstreaks are hardly ever seen feeding on flowers, as instead they seem to spend their time flying in short, swift bursts, chasing each other around bushes in secluded corners as they establish their territories. Its caterpillars feed on a variety of plants and shrubs such as *gorse*, *broom*, *common rock-rose*, *bird's-foot trefoil* and *dogwood*, so it is quite widespread although in small numbers.

As the summer encroaches, so the butterflies become more numerous. July and August with their preponderance of hot, sunny days is the time to see the beautiful *chalk-hill blue* butterfly on the grassy slopes of the reserve. The male is a bright silvery-blue with dark markings around the wings and is much more colourful than the darker and somewhat drab female. The colour difference is probably to help the male attract females, and to conceal the female from predators such as birds. Both sexes have a white edge to the wings and a row of spots on the hind-wings, orange in the female and black in the male. The chalk-hill blue is a local species found only where either the *horseshoe vetch*, *kidney vetch* or *bird's foot trefoil* grow, for it is on these plants that the eggs are laid in late August and on which the pale green hairy caterpillars emerge to feed some nine months later. If you have great patience, then find one of these caterpillars in April or May and watch it for a while with your hand lens. You may be lucky enough to witness a remarkable sight of mutual co-operation in the world of nature.

Watch for ants climbing onto the back of the caterpillar, not to attack it, but to stimulate a porous gland which then secretes a sweet fluid. This fluid, rich in nutrients, provides food for the ants. The caterpillar's reward comes later when it turns into a chrysalis in June. At the base of the food plant, the ants now construct a hideaway and place of safety for the chrysalis, made up of small pieces of chalk and other debris. They then keep guard over their 'charge' until it hatches out into an adult butterfly about a month later. The chalk-hill blue was once quite common on the South Downs, but the ploughing up of much of the grassland and the encroachment of scrub has meant a reduction of its food source, and thus of the butterfly itself. Luckily, the north escarpment of the South Downs here is too steep for ploughing, leaving scrub encroachment as the main threat to the butterfly. However, the advent of a nature reserve, with its associated resources management, has allowed this beautiful butterfly not only to survive but also to increase.

One of the birds you are almost certain to see on the Downs just about anywhere, but certainly here above Ditchling, is the *yellow-hammer*. You are likely to hear this before you see it, as the cock bird repeatedly sings a persistent little song from the top of a hedge, tree or bush, 'chi-chi-chi-chi-chi...chwee', usually described as 'little-bit-of-bread-and-no-cheese'. The male is easily distinguished by his yellow head and underparts, and chestnut upper surface, which are especially bright in the spring in order to attract a mate. The female is much less yellow, with darker markings particularly on the head. Both are about 16 centimetres long, and in flight the white on their outer tail feathers is particularly conspicuous. Their well-hidden nest is typically on the ground in a bank or at the base of a bush or hedge, and is built of grass and moss, lined with hair. The three to six eggs are laid from April to as late as August and are ashy-white and curiously marked with dark purple as if someone had scribbled all over them; this characteristic once gave the bird its country name of 'scribbling lark'. When autumn comes yellowhammers are to be seen in flocks in open fields feeding on seeds, worms or insects. In bad weather during winter months, they will join other hard-pressed birds to see what food can be had in and around farm buildings.

So on the hills above Ditchling look out for the yellowhammer as it sings its heart out on a spring or summer's day. As you take in the atmosphere of these hills your eye is drawn to light reflected off water. Near to the highest point at the beacon, and just to the east of the road, is a circular pond called a dew-pond.

These dew-ponds are scattered all over the South Downs and were constructed by man as a source of water for farm animals. Although downland turf made prime grazing for sheep and cattle, the underlying chalk, being pervious to water, meant that there were few, if any, natural sources of water on the Downs themselves. The origins and age of dew-ponds are surrounded in controversy and mystery; some people say they were built in the Saxon period, but others maintain they were started in the Middle Ages when the Downs were converted from mainly arable land to sheep pasture. Great skill was employed in constructing dew-ponds. First, a saucer-like depression about 2.5 metres deep, was scraped out and the resultant chalk spoil was used to build up the sides and to form a slight lip. The base was covered in straw, alternatively layered with clay brought up from the Weald. After several layers, about 6 centimetres of burnt lime was added to prevent worms puncturing the clay. The final layer consisted of a lining of locally gathered soil. Cattle and horses were prevented from entering the water, so possibly breaking the lining, by a suitably placed layer of flints. It is mist and rain, not dew, that maintains the water level in the ponds, and the efficiency of their construction can be seen in the large number that never dry out, even in the harshest drought. Modern dew-ponds, constructed since just before the Second World War, are made with concrete lining, but these have a tendency to crack in the frosts of winter.

Dew-ponds are marked on large-scale Ordnance Survey maps by small blue circles. Find one remotely situated, possibly near to a wood or scrubland and take time to sit quietly beside it. Wait patiently and watch, for your reward, especially in the dry months of midsummer, will be to see at close hand many birds and animals take life-giving sustenance from the pond. The ponds, now, are very much valued as important habitats for wildlife.

The views from the top of Ditchling Beacon are as extensive as they are impressive. The Weald, as always, lies beneath you to the north, buttressed further away by those other chalk hills, the North Downs. To the west is the great bastion of Wolstonbury Hill, described later in this chapter, and to the south are undulating sloping spurs, much cultivated now, just as they were thousands of years ago by Celtic farmers. On a clear day flashes of white catch your eye to the south-east; these are the chalk cliffs of Seaford Head and Haven Brow, one of the Seven Sisters. The bold outline of Firle Beacon stands out amongst a panorama of hills. Those above Kingston, a few miles distant, look interesting, and so retrace your steps to the area just south of Lewes and begin an exploration towards Brighton and beyond.

The track past Swanborough Manor, just south of Kingston, leads up to the crest of the magnificent stretch of downland that includes Kingston Hill, Swanborough Hill, Iford Hill and others. Some of the finest views of all the South Downs can be admired from these summits. The Ouse valley lies below; green with water-meadows interspersed with silver streaks of ditches, dykes, and little streams, all with their banks of lush vegetation. Mount Caburn and Beddingham Hill dominate this whole landscape of rolling, unenclosed earth whose summer sunburnt colours are seemingly all enshrouded with a vast expanse of blue sky. Along the valley lie little villages, Iford, Rodmell, Southease and Pidding-hoe. These villages, of Saxon origin, stand on the edge of the chalk, and once were riverside villages with quays and jetties, and their own fleet of fishing vessels. The Domesday Survey of 1086 indicates an extensive deep-sea fishing industry operating from the Ouse villages, with the villagers of Southease having to pay a rental of 38,500 herrings, and nearby Iford having to supply 16,000 herrings. At Piddinghoe there is a weather vane where a gilded fish swings to the wind, of which Kipling wrote:

> Or south where windy Piddinghoe's
> Begilded dolphin veers.

Perhaps this reflected the times when porpoises were seen regularly cavorting in the sea below as described in the last chapter; but perhaps not, as this fish seems only to be a salmon-trout!

But the sea was once supreme and held sway over the valley. It gave yet another industry to the villages along the edge; salt-making. Saltwater at high tide filled shallow pits and when the tide ebbed the water was held in the pits until it evaporated to leave salt. The Domesday Survey records 11 salt-pans at Rodmell and 4 at Beddingham. Salt was of vital importance to the mediaeval economy as a means of preserving meat; for farmers of those days had to slaughter many of their livestock each autumn because they lacked the means to provide winter fodder.

This stretch of downland to the west of the Ouse valley is well described in a poem by a Mrs Marriott Watson.

> Broad and bare to the skies,
> The great down country lies,
> Green in the glance of the sun,

Fresh with the clean, salt air:
Screaming the gulls rise from the fresh-turned mould,
Where the round bosom of the wind-swept wold
Slopes to the valley fair.

The bare downland slopes of these hills are disappointing for the botanist for although many of the typical chalk-loving plants can be found there, such as *marjoram, wild thyme* and *common rock-rose*, their numbers are somewhat sparse and you will find few, if any, of the less common plants such as *burnt orchid* and *early spider orchid*. In the late summer and early autumn, however, you should easily find the *autumn gentian* or, *felwort* as it is commonly called.

The *felwort* grows up to 10 centimetres in height, much taller than the *early gentian* described in Chapter 3, and it also flowers later, from July to September. Its base leaves are spoon-shaped, whereas those on the stem are lance-shaped and form opposite each other. The lipstick-like buds are borne in small clusters up the much branched stem and open out into four or five-petalled, dull purple flowers.

Another member of the gentian family which you will find all over downland slopes and also here above the Ouse valley is *yellow-wort*. Up to 40 centimetres tall, yellow-wort has a rosette of spoon-shaped leaves at its base from which the very erect round stem grows. The stem leaves are in pairs with their bases joined so that they completely surround the stem, making it difficult for crawling insects to climb up to the flower. The bright yellow flowers appear in June and bloom until September and are pollinated by flying insects. The whole plant has a glaucous hue which easily distinguishes it from others of the same family.

Before making your way westwards across country on well-marked tracks and through cultivated fields, pause for a while at Breaky Bottom just west of Rodmell. Here, tucked away in a seemingly remote valley, are rows and rows of vines, neatly trained along wires stretched between wooden poles. This small, six-acre vineyard produces prize-winning, fine dry wines and is open to the public all the year round. When you descend into the valley itself you will notice the change in atmosphere; for here, because of the geography, this little downland combe has a micro-climate all of its own. The air is warmer and the mood and quality of its surroundings give it an ambience that is unique. Do visit and taste some of its wines.

Balsdean, and more importantly, the Castle Hill Nature Reserve, is your next goal with magnificent vistas all about you on these high downland paths. If you look to the south you will not fail to see lines of houses, ugly in their conception which mar the view of a distant sparkling sea. It was back in the 1920s that man set about despoiling these places of peace. Walk these lonely tracks in the spring or autumn and you might just be lucky enough to see one of our scarce migrant plovers, the *dotterel*.

About 21 centimetres long, the *dotterel* can be distinguished by the white band between its orange-chestnut underparts and its blackish crown with broad white eye-stripes. In flight, its white face, dark underparts, russet flanks and white breast-bar, are quite distinct. It breeds on the high tops of Scandinavian and Scottish mountains and migrates southwards to the Mediterranean and Africa in October, only to return in April and May. On the way northwards in the spring, the dotterel feeds on anything it can find amongst the growing crops on downland hills, particularly beetles, flies, and spiders. Strangely the Downs above Kingston and Balsdean Farm are much frequented by migrating dotterels and some fields here are known locally as 'dotterel fields'. Certainly many of the Sussex sightings over the past 40 years have been made on the Downs between Woodingdean and Rodmell. Interestingly, the female is slightly larger and more handsomely coloured than the male, and after laying her three eggs amongst short heather and lichens on a remote mountain top, she leaves it to the male to incubate them and to raise the young. Meanwhile she goes off to search for another mate and to repeat the process, perhaps another two or three times. In their breeding locations dotterels are very tame and have been known to sit tight on their nests while having their backs stroked.

Balsdean, the 'dean' or 'dene' being a Saxon word for valley, is a collection of modernised farm buildings situated on High Hill above a relatively wide, flat-floored valley which was once the watercourse of an ancient stream. Near the bottom of the valley is a water pumping station and close by, there is just a hint of ancient dwelling places set alongside the broad path contouring the dene itself. Perhaps long ago the sea penetrated this place and its inhabitants may have complemented their tilling of the land with fishing. Balsdean Farm has its own interesting history. In the early Middle Ages of the twelfth and thirteenth centuries, chapels were built in isolated hamlets so that hard-working peasant farmers could worship without the need for time-consuming travel to a distant parish church. Such a chapel was built at Balsdean, later converted into a stable in the nineteenth century, and then incorporated into the farm. Now, there is no sign of that ancient place of worship.

The trackways around Balsdean are extensive; walk upon them on a day in high summer. Little scattered barns are set in acres of golden brown wheat, and all around there is an ambience of quality where only the distant calls of scavenging *rooks* break the sun-soaked silence of solitary downs. Here, once again, solitude is only a kilometre or so from the suburban sprawl of seaside towns like Saltdean and Peacehaven. *Swifts* gather, wheeling and swooping through the hot humid air of a sultry summer day, catching flying insects and fattening up, ready to make their great journey across the seas to Africa.

Beside these dry tracks around Balsdean Farm are many wild flowers, *common toadflax*, *greater knapweed*, *wild mignonette* and *common mallow*; all giving a kaleidoscope of colour to an impressive scene. In the distance the steep-sided slopes of Castle Hill Nature Reserve come into view, but before exploring this area it is worth a short diversion to the main A259 road at Rottingdean. On the seaward side of this busy highway from July to September can be seen the purple spikes of one of our rare species of *sea-lavender* which are familiar wild flowers around our coasts. This particular species is one of the rock sea-lavenders which grows wild nowhere else in Britain and has been named the *Rottingdean sea-lavender*. It differs from the *common sea-lavender* which thrives on salt-marshes, by being smaller, some 20 centimetres tall as opposed to 30 centimetres, and being neater and more compact and upright, having a stem that branches from near the base. It has lilac and bluish-purple flowers and lance-shaped leaves. Interestingly this variety of the rock sea-lavender comes from the shores of Sicily and has probably escaped from a garden rockery. See if you can spot it growing amongst the grass on the banks and edges of the busy main road; this seaside plant looks somewhat incongruous on top of these chalk downland cliffs.

Retrace your steps, past Balsdean Farm and on to Castle Hill Nature Reserve: once this was the site of an Iron Age hill fort which has long since ceased to exist. The reserve, with an area of 280 acres, is one of the many National Nature Reserves in England which is managed by English Nature. Essentially it consists of unploughed downland slopes with chalk grassland and areas of scrub. Here is one of the best examples in East Sussex of the nationally uncommon, chalk grassland habitat. This site provides a rich variety of flowering plants and grasses, and of course, butterflies and insects. Much of the scrub is dominated by *gorse*, *hawthorn*, *wayfaring tree* and *blackthorn* which supports breeding birds such as the *stonechat*, *linnet* and *whitethroat*, amongst others. Typical of the many wild flowers here are *horseshoe vetch*, *kidney vetch*, *yellow rattle* and the rare *Nottingham catchfly*.

The *horseshoe vetch* is common on the reserve, flowering from May to August. Its groups of six to ten bright yellow flowers held at the end of slender stalks, some 3–5 centimetres long, show up well amidst the downland grasses. It could be confused with a similar, and much commoner plant, the *bird's-foot trefoil*, but the horseshoe vetch has quite distinctive leaves, consisting of three to eight pairs of oval leaflets each held on a single stem. In the summer it is easily recognised by its peculiar snake-like pods of many one-seeded joints, each curved in a horseshoe shape from which the plant takes its name. Its Latin name, *hippocrepis*, is derived from two Greek words, *hippos*, a horse, and *krepis*, a shoe.

The *kidney vetch* has bluish, often hairy leaves made up of three to nine lance-shaped leaflets. The pretty yellow, sometimes creamy-white or even crimson flowers, are held close together, usually in bunches of two at the end of 10–30 centimetre-long stems. The flowers bloom from June to September and are set in a dense, woolly coating of silky hairs. These were found useful in early times for staunching blood and hence its Latin name *vulneraria* meaning a cure for wounds. The English name is thought to refer to the shape of the flower heads like a two-lobed kidney. Another name for the plant is lady's fingers from the shape of each individual flower.

Both the kidney and horseshoe vetch are food plants for many species of butterfly larvae, particularly of the *small blue* in respect of the former, and the *chalk-hill* and *Adonis blue* for the latter; therefore these butterflies thrive well on the reserve.

The *yellow rattle*, another characteristic plant on the downland slopes, grows up to 40 centimetres tall with a straight stem that terminates in a cluster of yellow flowers; these bloom from May until August. Each flower has an upper hood and a lower lip and after flowering it develops into a swollen, bladder-like capsule, which contains numerous seeds. When the plant is shaken, these seeds make a distinct rattling, hence its name. Yellow rattle is a semi-parasite, its roots drawing food from neighbouring grasses and herbs.

A very rare member of the Sussex flora that you may find on the Castle Hill Reserve, is the *Nottingham catchfly*. This attractive plant, up to 60 centimetres high, has white or pink flowers about a centimetre wide, which have narrow deeply-cut, two-lobed petals. The evening-scented flowers appear from May to July and the petals only unroll at dusk to attract night-flying moths for pollination. The plant has downy, somewhat sticky stems and broad lance-shaped leaves. The Latin name, *nutans*, refers to the nodding flowers, while the English name recalls an early record on the walls of Nottingham Castle where, sadly, it no longer occurs.

Many other wildflowers grow in this rich habitat. In May and June, *dropwort, salad burnet,* and several *vetches* colour the slopes, and in July and August *yellow-wort, common centaury,* and *small scabious* are most noticeable. Some less common plants which have been mentioned in previous chapters may also be found, such as the *early spider orchid, burnt orchid* and *field fleawort*. All the colourful blooms of the downland flowers tend to make us forget that the real matrix of the turf is grass, so it is worth giving a brief description of just a few grasses, like *sheep's fescue, common quaking-grass, upright brome* and *tor grass.*

Sheep's fescue forms thick tufts, 5–60 centimetres tall. The leaves are thread-like and the spikelets of flowers are a glaucous green with a violet tinge. It is abundant on hill pastures where it is a valuable diet of sheep.

Common quaking-grass is found throughout Britain on most types of grassland and here on the South Downs it is one of the most attractive of its kind, the flower heads often being picked and dried for winter flower arrangements. It is about 20–50 centimetres high and the spikelets of purplish and drooping flowers shake with the wind on slender stalks. It flowers from May to August.

The commonest tall grass of the chalk downland with stems some 40–80 centimetres tall is *upright brome*. Its oblong spikelets of purplish flowers are held upright on their stems and appear from May to July.

Tor grass, or *chalk false brome* as it is sometimes called, grows 30–60 centimetres tall, has stiff conspicuously yellowish-green leaves, and spikelets of upright flowers held close to the stem. The flowers appear from June to August. Tor grass with its creeping root stems forms large, yellow-green, circular patches all over the Downs. Unfortunately it is so coarse that it is eaten by only the hungriest of grazing animals and thus creates a dense tract of grass that smothers most other plants. It is therefore, a problem, particularly as it is very difficult to eradicate.

This has been but a brief mention of a few of the many, many species of downland grasses. It would help if you use one of the field guides on grasses to identify and learn more about them.

Interestingly, here on Castle Hill Nature Reserve, research is being conducted on the effect of grazing. Areas of ungrazed chalk grassland are enclosed in wire fencing and it is easy to see the effect of grazing. When I visited the reserve in May 1999, the grass inside the fence was some 50 centimetres high with a profusion of tall flowering plants. Outside the fence where the sward had been grazed, it was barely 5 centimetres tall and contained only a few prominent flowers, mainly of the smaller variety like *dove's-foot crane's-bill* and *fairy flax*. The local and uncommon *field fleawort* could be found on the reserve in large numbers, but sadly, many had been eaten and the flower stems of those that remained were short and bedraggled. As on Mount Caburn, the grazing, to my mind, had been excessive. I look back with nostalgia to the time of my boyhood and can remember the downland slopes covered with grasses as long as that inside the wire enclosure on Castle Hill. There were many more butterflies, bees and other insects about then, and one could see the ripples of the wind across the greens-ward and watch the sway of flowers with their multitude of colours. How often in days gone by have I sat amongst the long grass on a summer afternoon and watched the cloud shadows sliding with the wind across hills and woods. Patches of sunlight would suddenly appear bringing unexpected brightness to fields and hedgerows. There are not many places now where one can experience such delights on our over-grazed Downs

The rich variety of grasses and wildflowers are valuable for grasshoppers and crickets, an invertebrate group for which the reserve is possibly of national importance. As well as the *great green bush cricket*, described earlier, the reserve is the site for the rare *wart-biter grasshopper*. This large green insect, some 4 centimetres long, differs from the similar great green bush cricket by having shorter wings which have prominent black dots on them. The wart-biter, once used to remove warts, has particular requirements; a mixture of long grass for concealment, and also short turf where the sun's rays can give it the necessary warmth. Gaps in the turf are also required for the female, in which to lay her eggs. These essential requirements can all be met on the reserve and so the species continues to thrive.

Butterflies abound on the reserve, and one of the specialities here is the *small blue*. As Britain's smallest butterfly with a wingspan of only 22 millimetres, this species of blue is easily over-looked, but can be seen best in June when the males can be found basking on long grass with wings outstretched awaiting passing females. Both sexes are sooty-brown in colour on the upper sides of their wings with the male having a tinge of blue. Both sexes also have silver-grey undersides with black dots and blue at the base of the wings. The small blue is a vigorous flier and skims low over the grasses and flowers, sometimes flying through thick vegetation where its colouring gives it good camouflage. The constant movement through plants gradually wears away the wing scales and, as a result, they soon lose their colour, their wings become ragged and they lose the ability to fly. After about two weeks of life they fall to the ground to be eaten by ants and birds. The female lays her eggs singly in the flower heads of *kidney vetch* which hatch out into small brown caterpillars a week later. In July, when fully grown, the caterpillars spin the flower heads together over themselves and hibernate until the following spring. They turn into chrysalids in May and the adult butterflies emerge two weeks later. Interestingly, if one small blue caterpillar encounters another on the same flower, it will attack it, and the stronger will eat the weaker. Such an occurrence is not uncommon in the cruel reality of nature.

One of the characteristic birds of the gorse scrub that covers some of the slopes, is the *stonechat*. In April and May you will possibly hear its persistent scolding call, 'wheet, tsack, tsack,' like two stones being knocked together, before you see the bird itself. The male is easily distinguished by his black head and white patches on the side of his neck, features which are absent from his mate. They like to perch on the top branches of small bushes, especially gorse, and they feed almost entirely on insects and grubs. Their nests are carefully concealed amongst grass, brambles, or at the foot of gorse bushes, and are made of grass and moss, with a lining of hair, feathers and fine grass. Although some stonechats migrate southwards in the autumn to Spain and North Africa, most overwinter near the coast and so can be seen all the year round.

On the way to Hollingbury Hill just to the north of Brighton, stop for a while and enjoy the atmosphere of Falmer pond, hidden and tucked away from the busy A27, in what was once a fairly isolated Sussex village. Here, beside the village green and church, is a small area of placid water with an island in the middle, so sit on the grass around it and enjoy its serenity. On the north side of the pond, note the collection of about 100 sarsen stones which were collected up in the early 19th century from the Earl of Chichester's land nearby. They were brought at much trouble and expense and placed about the pond edge and around the ancient village pump. Some of the sarsens are of huge size and weight and form the largest collection of such stones in East and West Sussex. Sarsen stones are blocks of sandstone roughly rectangular in shape. They are thought to be residual boulders from a bed of sandstone which once covered the chalk and were eroded into their present shapes by centuries of weathering. However, some believe the stones were carried south from other parts of the country by ice age glaciers and deposited over the South Downs. Whatever their origin, they are found all over Southern England from Wiltshire to Kent. Many have been moved by man who used them for building, and for marking boundaries or religious sites. The name sarsen is derived from the seventeenth-century use of the word saracen to denote something unusual and foreign. These stones are also known as 'grey wethers' because of their resemblance to sheep lying on the ground.

As you sit on the grassy bank on a warm spring day, watch the *moorhen* gathering small green stems and dead water plants to build up its platform-shaped nest in the vegetation surrounding the island. The moorhen, about 32 centimetres long, is easily distinguished by its dark plumage and prominent red forehead and white stripe along its flank. It is found on most ponds and areas of water which it shares sometimes with a larger member of the same family, the *coot*. The coot, however, has a white blob on its forehead. The moorhen lays seven to twelve buff-coloured eggs, spotted and speckled with reddish-brown, from March to as late as August. They hatch out after three weeks and the young leave the nest in two to three days as little bundles of black fur. The moorhen, like the coot, feeds on just about anything it can find, from slugs and worms to insects, seeds and waterweed, and of course it often feeds off scraps thrown by people into ponds and lakes. Sometimes the moorhen will devour its eggshells after hatching, to retain the necessary calcium and mineral salts in its body. A poor flier, the moorhen is truly at home in the water where it spends the whole year near thick undergrowth in which it can hide if danger threatens. It can remain submerged for quite long periods with just its bill above water. If disturbed at close quarters from a quiet country pond or stream, it will rise from the water with a pattering of feet along the surface, uttering a harsh penetrating 'kittock'.

Tall *sycamores* with their masses of matt-green foliage and greenish-yellow, drooping flowers give a sense of space to this pleasant spot. Sycamore wood, being of 'clean appearance', cream-coloured, heavy and fairly hard, is used in making furniture, textile rollers, and bowls, spoons and platters. If the wood has a rippled grain it fetches a high price for veneers. Young *English elms* also grow around the pond, seemingly unblighted by the Dutch elm disease that has greatly reduced the numbers of this fine tree. The English elm, before the outbreak of this disease in the early 1970s, was a common sight along our hedgerows, where it was planted deliberately by eighteenth century landowners. When their tenant farmers began to enclose the old open fields with hedges of hawthorn, the landowners required them to plant elms at intervals for shade, shelter and timber. These elms were raised from nursery stock, and when the mature trees were felled they were replaced automatically by the sucker shoots that spring vigorously from the roots. Dutch elm disease is caused by a beetle burrowing into the tree and spreading a fungus that destroys the wood tissue just below the bark. Many of the great elms of Southern England and South Wales have died as a result of the disease and great efforts are now being made to eradicate it and prevent its spread. Sadly, our landscape once characterised by tall, elegant English elms has changed dramatically with their loss. The heartwood of the English elm is reddish in colour, firm and heavy, and, because it has an interlocking grain, it is difficult to split and is long-lasting even if kept continuously wet. Among its uses are coffin boards, chairs, tables, garden furniture and underwater goods; an important asset to any country. Let us hope these elms around Falmer pond will grow into the tall graceful trees that were once so familiar to us all.

Amongst the crowded vegetation growing around the margins of the pond, see if you can find *gipsywort*, a tall plant which can grow up to almost a metre tall. It is the large elliptical, deeply cut, saw-toothed leaves that are likely to attract your attention first. They grow opposite each other on the square stem, and where the upper leaves join the stem, there you will see the small whorls of whitish flowers from June to September. Interestingly the juice of the plant has long been used to give a fast black dye which was supposed to have been used by gipsies to darken their skin, hence the name.

Other plants you will find amongst the grass around the pond are *spotted medick* and *scentless mayweed*. Medicks are one of several groups of plants belonging to the pea family whose leaves are divided into three similar leaflets. The small, yellow, pea-like flowers appear from May to August, and the whole plant is prostrate on the ground, with creeping and slightly downy leaves. The spotted medick species is easily recognised on account of the dark spot on each leaflet.

The *scentless mayweed*, this time a member of the daisy or

composite family, is another plant that spreads itself across waste ground. It has scentless flower heads, some 2–5 centimetres across, with white petals and a central dome of yellow, tubular florets. It has numerous soft, green branches, well covered with alternately spaced, stalkless and finely cut leaves. This is one of our commonest daisy-like arable weeds so you should easily spot it flowering from May to October.

Time to move on from this picturesque place and visit an intriguing hill just to the north of Brighton, Hollingbury Hill. I chose to visit this evocative place early one June morning to capture the dawn and to watch the sun rise, phantom-like, over distant downs. Hollingbury Hill is another of the Iron Age downland forts built around 450–400 BC. Its rectangular ramparts on the top of the hill enclose an area of some nine acres. Once this place was the site of inter-tribal wars but now there is only a golf course and a nature reserve; so peace has once again returned to its surroundings. However, Brighton with its monotonous vista of houses continues to encircle this place and there remains only a narrow strip of green to the north to remind one of an age when this was all bare downland and, the haunt of sheep and the shepherd.

Approaching Hollingbury Hill from the north, an ever lightening sky shows you the path from the car park to the summit. The ancient ramparts become visible as you ascend the hill, mysterious, and then sinister, as you spot the dim silhouette of a person standing alone, looking slightly menacing. Unmown grasses and many wild flowers lie at your feet as you approach the top and the inevitable encounter with the stranger. He is an agreeable young man intent only on watching the sunrise as he explains to you the mysterious lines of energy, ley lines that converge on this spot. These lines are marked by enigmatic and strange stones placed upright by early man all over the country. Here, at Hollingbury the ley line passes to the south through the great sarsen stones around the Dolphin Fountain in Brighton Old Steine. The Steine is the valley of a lost river, the Whalesbourne; steine or stein is of course an ancient word for stone. Curiously, further south the line passes through St Nicholas's Church in Dyke Road, where a lingering tradition has it that there was once a circle of stones here which formed part of a 'druidical circle'. There is a plinth of sarsens still in place just below the south entrance of the church.

To the north, the line passes through Peckers, a local name given to a place near Upper Lodge Wood to the west of Stanmer, where archaeologists are conducting an excavation of an ancient site. There were almost certainly standing stones, now recumbent at

106

Peckers and much more will no doubt be discovered about the early history and meaning of the stones. Nearby Standean, meaning stone dean, was once a valley of sarsen stones, now all gone, gathered and swept off for building purposes.

Whatever unknown forces give us ley lines, the stones that seem related to them are real enough. Solitary upright stones, or stone circles, are something of an unknown world to man; a lost religion perhaps, whose origin is now just a memorial of something gone from human recollection.

Gazing into the heavens and watching the red ball of the sun appear over distant Blackcap, and in a rising ellipse pass over Mount Caburn, you ask yourself the silent question: 'Where did it all come from?' or even, 'Are we alone?'

Within the past 20 years much more has been discovered about our universe. Mathematicians and astronomers have calculated that the 'Big Bang' origin of the Universe was not just any old bang but an explosion of meticulously arranged magnitude. Indeed, if a change were made to the strength of gravity inside a star of only one part in 10-to-the-power-of-40 (i.e. 10 with 40 noughts after it) then this would be catastrophic for a star like the sun. So the development of life on planets is made possible only by very fine adjustments to huge numbers at the time our universe was created. This is known as the Anthropic Principle and states that at the instant of 'Big Bang' the Universe was immediately fine-tuned to guarantee the eventual emergence of intelligence. This principle, with its irresistible logic, points so plainly to the existence of a Creator, that agnostics have cast around for other theories to repudiate it, but with little success.

All very interesting as you look down at Brighton where street lights shine out in the gentle early morning mists of pre-dawn. This great metropolis of man is hushed and here at the top around the concrete triangulation point nature is still dominant. A *blackcap* sings a first song of morning from the top of a hawthorn bush. The blackcap has a remarkably rich, warbling song and when alarmed responds with a rapidly repeated 'tac, tac' and a harsh churring. About 14 centimetres long with a sober grey plumage, it owes its name to the glossy black cap of the male, the female's being more of a bright brown than black. The blackcap haunts bushy situations where briars, thick hedges, thickets and nettles provide a suitable site for its fragile nest of grass, lined with hair. Here on Hollingbury Hill, the nature reserve provides just what it needs, downland slopes covered with dense scrub and woodland, consisting mainly of *field maple*, *ash*, *whitebeam*, and *hawthorn*. The blackcap is a migratory bird that arrives from Africa in late March to early May. Although a great insect-eater, the blackcap has a taste for fruit and berries and this no doubt enables many, mostly males, to live through the winter in this country without moving south to warmer climes. Often a male overwinters in the small orchard and tall hedges of my own suburban garden.

As the sky lightens, two conspicuous plants catch your eye, *dropwort* and *lady's bedstraw*. The dropwort is widespread on chalk or limestone grasslands and is sometimes referred to as 'the meadowsweet of the Downs'. Indeed, it is a similar plant to the much taller meadowsweet of our damp meadows and roadside ditches, but its creamy-white elegant flowers are not fragrant and its leaves are more jagged and deeply divided and form a rosette around the base. It is about 20–40 centimetres high and flowers from July to September.

Lady's bedstraw, with its large compact clusters of small yellow flowers, adds colour to the grassy slopes of the Downs in summer. It has wiry, much branched stems, some 30 centimetres long, which bear whorls of 8 to 12 slender leaves along their length. In places, where its foliage straggles through the grasses, it forms quite large colonies which are easily seen and recognised. Lady's bedstraw, when dried, gives off a pleasant scent of new-mown hay and many years ago was included in straw mattresses, especially in the beds of women about to give birth; hence its name. The flowers were also used to curdle milk when making cheese. In Gloucestershire it was mixed with the juice of nettles to make Double Gloucester cheese, and it was also used to colour Cheshire cheese. The flower tops, distilled with water, produce an acid liquid which can be mixed with other juices to form a pleasant summer drink. Lady's bedstraw, a plant for all seasons if ever there was one, was also once used in dye-making.

As you descend the slope and think about your next destination, Wolstonbury Hill, the discordant noise of early traffic building up on the A27 intrudes on your thoughts and feelings and brings harshness and reality back to life.

As you make your way towards Wolstonbury Hill by the A27 bypass and the A23, you cannot fail to see the extensive array of wildflowers flourishing on the embankments created by new road schemes. Such places now provide a rich habitat for wildlife of all kinds. Here is a haven for the *small blue* butterfly, attracted by the abundant kidney vetch. The *kestrel*, too, is a common sight, seemingly oblivious of the pounding noise of traffic as it hovers over the rich sward seeking out small rodents.

If you are not walking along the South Downs Way, then the best access to Wolstonbury Hill is from the north. Park your car in a small lay-by beside the lane that runs south from the B2116. Then take the path that leads through the wood just south of a private house, 'The Warenne.' As you walk along this sunken path beneath the shade of tall trees, look for the *white helleborine* and the *fly orchid* described earlier. Another orchid that you may find growing in shady places beside the path is the *greater butterfly-orchid*. A fairly tall orchid with a single stem 20–60 centimetres high, the orchid is most attractive, with its large white flowers showing up well, especially in the early evening, when they put out their strong sweet scent to attract night-flying moths. Each stem bears some 10–25 flowers which come into bloom at the end of May and last until the end of June. At the base of the stem are two large pale green oval leaves, above which are several small, pointed stem leaves. You will find this uncommon orchid in several woods below the South Downs where the soil is heavy, damp, and moss covered. About a kilometre up the path you will see a gate and a stile on your right with open downland beyond. The path here leads to the summit of Wolstonbury Hill.

As you walk up the gentle slope typical downland plants greet you, such as *yellow-wort* and *common rock-rose*, and on the steep sides of ancient chalk workings, you will see purple patches of *wild thyme*. One attractive plant you will not fail to see here growing on grassy banks, is the *germander speedwell*.

Flowering from March to July, the *germander speedwell* has intensely bright blue, very attractive flowers with darker blue veins and white centres. Its stems are about 30 centimetres long and creep along the ground and then ascend. The stems have two opposite rows of long white hairs and the oval bluntly-toothed leaves are also covered with hairs. The germander speedwell is also known as cat's-eye and birds-eye speedwell, and many still call it, erroneously, forget-me-not. There are over 20 species of speedwell growing in Britain, so it is worth using a good flora to check on the subtle differences between some of the commoner ones. Many speedwells are to be found beside trackways and roads and were so named 'to speed you on your journey'. Nobody is sure why they acquired this reputation, but in Ireland they were sometimes sewn into clothes of travellers for good luck.

As you approach the summit of Wolstonbury with its remnants of an important Iron Age fort, so the views begin to unfold, unimpeded by tree or building. From here the dim blue plains and hills of the Weald can be seen, and all around the distant ridges, the wide expanses and bare folds give a primitive changelessness to these heights. However, the sense of peace and remoteness from everyday life is marred yet again by the ceaseless drone of heavy traffic on the A23 which carves its way below through a fold in the escarpment.

As you examine the hill-fort itself, you will notice that the usual protective ditch has been dug on the inside of the rampart instead of the outside. One would have thought that such a design would much reduce the defensive value of the fort itself. You will also see around the summit trig point, the remains of nineteenth-century archaeological diggings, now covered with turf. These have produced numerous dips and hollows which are useful as a shelter against the wind.

On the grassy slopes of this beautifully formed hill, now protected from human interference by the National Trust, you may be fortunate to find the rare *man orchid* in one of its few sites in Sussex; so called, because the greenish-yellow lips of each flower strongly resembles the outline of a man, complete with arms and legs. The outermost part of the flower forms a helmet or hood for the 'man's' head. The flowering spike bears as many as 90 yellow-ish-brown flowers at the top of a stout stem, 15–40 centimetres high. The long flowering period is from May well into July. At the base of the stem there are three to four broad, lance-shaped, bluish-green leaves which have prominent veins. The upper leaves are much narrower and clasp the stem and often appear scorched at their tips.

Another remarkably attractive and striking orchid, the *bee orchid*, can also be found amongst the grass on top of Wolstonbury Hill. This is more common than the man orchid and is more wide-spread on the South Downs. It flowers slightly later, from June to mid-July, and is easily recognised because of the flower's resem-blance to a bee. The flower stem ranges in height from 15–40 centi-metres and at its base are five or six greyish-green, oblong leaves; by mid-summer, as with the man orchid, these are often burnt brown at their tips. The upper leaves are narrower and clasp the stem. The flowers themselves are well separated and are each made up of beautiful colours, brown, yellow, and pink all put together to look like a bee. It is not altogether certain why this mimicry should occur; most probably the idea is to attract bees to mate with the look-alike and thus ensure pollination. Interestingly, the bee orchid flowers are also self-pollinating, the pollen masses situated so that they can fall straight on to the female organs. Occasionally the bee orchid will hybridise with other orchids such as the *fly orchid* to

Plate 1   Firle Beacon, Page 55   Painting by Frank Wootton, OBE

Plate 2   Beachy Head and Lighthouse, Page 27   Author

Plate 3   The Fulmar on the Seven Sisters, Page 31   Author

Plate 4   Mist over the Seven Sisters, Page 31   P.D.L. Maurice

Plate 5   Alexanders in Eastbourne, Page 24
P.D.L. Maurice

Plate 6   Broad-leaved Helleborine, Page 36
P.D.L. Maurice

Plate 7   Firle Beacon across the Cuckmere Valley, Page 48   Author

Plate 8   Cowslips near Belle Tout, Page 30   Author

Plate 9   Field Fleawort, Page 28   Author   Plate 10   Early Spider Orchid, Page 29 Author

Plate 11   Oilseed Rape, Lullington,   Page 48
P.D.L. Maurice

Plate 12   Rough Marsh Mallow,   Page 49
Author

Plate 13   Common Poppy at the field edges, Page 49   Author

Plate 14   The Seven Sisters from Hope Gap, Page 64   Author

Plate 15   Sea Pink at Seaford Head, Page 63   Author

Plate 16   Clouded Yellow (Var Helice), Page 70   Author

Plate 17   Rock Samphire at Tide Mills, Page 68   D.L. Vinall

Plate 18   Rook at nest, Page 46   M. Hollings

Plate 19   Sainfoin, Page 72   D.L. Vinall

Plate 20   Mount Caburn in July 1983, Page 77   Author

Plate 21   Mount Caburn heavily grazed in July 1999, Page 77   Author

Plate 22   Malling Down, Page 81   Author

Plate 23   Devil's Dyke and Saddlescombe Chalk Pit, Page 114   M. Hollings

Plate 24   Chalk-hill Blue, Page 77   Author

Plate 25 Musk Orchid, Page 79   Author   Plate 26   Silver-spotted Skipper, Page 80
                                                                          Author

Plate 27 Man Orchid, Page 110   Author   Plate 28 Bee Orchid, Page 110   Author

Plate 29   The Ouse north of Lewes, Page 82   Author

Plate 30   Female Black Cap, Page 107   Plate 31 Childing Pink, Page 119   Author
                            M.Smith

Plate 32   Ditchling Beacon looking west, Page 90   B. Meldrum

Plate 33   Jack and Jill Windmills looking east from Wolstonbury Hill,
                Page 112   Author

Plate 34   Looking east to Newtimber Hill from below Truleigh Hill,
                Page 112   B. Meldrum

Plate 35 Burnt Orchid, Page 77   Author   Plate 36 Meadow Clary, Page 116   Author

Plate 37   Sheep on Castle Hill Nature Reserve, Page 102   Author

Plate 38   Shoreham Beach with Erigeron and Sea Kale, Page 120   Author

Plate 39   Shoreham Beach with Sea Kale, Page 119   Author

Plate 40   Starry Clover in flower, Page 121   Author

Plate 41   Starry Clover in fruit, Page 121   M. Hollings

Plate 42   Chanctonbury Ring, Page 125   M. Hollings

Plate 43   Amberley Wild Brooks, Page 133   B. Meldrum

Plate 44   Lapwing on nest, Page 134   M. Hollings

Plate 45   Male Yellow Wagtail at nest, Page 136   M. Hollings

Plate 46   Wallflowers on Amberley Castle, Page 140   Author

Plate 47   Broad-bodied Libellula, Page 138   M. Hollings

Plate 48   Path north of Burpham, Page 143   Plate 49 Moth Mullein, Page 53   Author
                                     Author

Plate 50   Arundel Castle from near Burpham, Page 149   Author

Plate 51   Rewell Wood, Yew Tree Gate, Page 155   M. Hollings

Plate 52   Pearl-Bordered Fritillary, Page 157   M. Hollings

Plate 53 Elecampane, Page 74 Author Plate 54 Autumn Lady's-tresses, Page 142

Author

Plate 55 Harting Downs, Page 198 B. Meldrum

Plate 56   Male Treecreeper, Page 204   Plate 57   Spotted Flycatcher, Page198

M. Smith   M. Smith

Plate 58   Duke of Burgundy Fritillary, Page 197   M. Hollings

Plate 59   Little Ringed Plover, Page 152   M. Hollings

Plate 60 Fly Orchid, Page 85   Author   Plate 61   Greater Butterfly-orchid, Page 109
                                                                                 Author

Plate 62 Willow Warbler, Page 146   M. Smith   Plate 63 Dog Rose, Page 153   M. Smith

Plate 64   Halnaker Windmill, Page 177   Author

Plate 65 Red Helleborine, Page 205   Author   Plate 66 Narrow-leaved Helleborine, Page 205
Author

Plate 67   Stone Curlews at nest, Page 178   M. Hollings

Plate 68   Canada Geese, Page 169   Author

Plate 69   Song Thrush at nest, Page 188   M. Hollings

Plate 70   Jays at nest, Page 204   M. Hollings

Plate 71   Male Orange Tip Butterfly, Page 175   M. Smith

Plate 72   Female Whitethroat at nest, Page 75   M. Smith

Plate 73   Fieldfare, Page 188   M. Hollings

Plate 74   Pagham Harbour, Page 181   M. Hollings

Plate 75   Little Tern at nest, Page 183   M. Hollings

Plate 76   Downland Track in Winter, Page 224

Painting by Christopher Osborne

Plate 77   St Andrew's Church, Alfriston, 'The Cathedral of the Downs',
                Page 224   Painting by Christopher Osborne

form a truly unusual and impressive flower. Other varieties of the bee orchid exist, such as one with white replacing the pink flower parts, at the same time having a lip that appears light green, instead of brown. Another has a lip coloured pink with the central of the three lobes being quite narrow and pointed. So keep your eyes open and look out for some of these curious varieties; they could appear anywhere on the Downs.

While sitting and contemplating the views all around, your eye catches sight of a moth flying fast over the grass. It settles on a low bramble bush and you notice its four prominent false eyes, one on each wing. It is an *emperor moth*, the only British member of the silk-moth family but its silk is not suitable for commercial purposes. This one is a male which is smaller and has a brighter colour than the female, with orange hind-wings and prominent feather-like antennae. The female has grey wings about 8 centimetres across and, unlike the male, does not fly during daylight. The male emperor moth has a well developed sense of smell and uses its antennae to detect females sometimes as far away as several hundred metres. The caterpillars feed mostly on brambles and heather and they spin their cocoon of silk in July and August in the foliage of the food plant. The adults emerge in the following April and May, which is the time you will see them.

High above you a *skylark* sings and appears almost to touch the slow white sails of small clouds which drift with the wind causing shadows to sweep across the fields below. Over these fields small birds appear, flying some 30 metres off the ground, diving and skimming around some farm outbuildings. With binoculars you notice their white underparts and distinct white rump near the tail. These are *house martins* which arrived only a month ago, at the end of April, from the middle of Africa. They are clearly feeding on flying insects and soon they will be gathering material to build their nests of mud, lined with straw, close up under the eaves of the farm buildings. The four or five white eggs are usually laid in May or June, the young hatching out and maturing well before they return to Africa in September or October. Sir John Squire admirably describes them in his poem, 'The Birds':

> And, skimming fork-tailed in the evening air,
> Where man first was were not the martins there?
> Did not those birds some human shelter crave,
> And stow beneath the cornice of his cave
> Their dry tight cups of clay? And from each door
> Peeped on a morning wiseheads three or four.

To the east there is a fine view towards those two well-known windmills, Jack and Jill, which side by side are an outstanding landmark along this range of the South Downs. Jack was built in 1876 and worked until 1908. Jill was built in 1821 in Brighton and was towed by a team of oxen in the middle of the century to her present site where she worked until 1909. Both mills used the power of the wind to grind corn. A distinctive chalk-white trackway winds its way over the hills and past the windmills. Take a walk along this route and see the tall *twayblade orchids* amongst others that grow in the woods just below. Both the *whitethroat* and *lesser whitethroat* may be heard and seen amongst the bushes and undergrowth nearby, together with many other birds such as the *linnet* and *goldfinch*.

If you look to the south and west from Wolstonbury Hill across the broad ribbon of highway and fast moving traffic, you will see our next port of call, the tree-covered Newtimber Hill and the rolling bare downs above Poynings and Fulking.

Newtimber Hill and its associated wood and chalk pit is best approached from the north along a track leading off from the A281. May and June are good months to visit this National Trust site as the magnificent trees of *ash* and *beech* are in full leaf then and the woodland orchids such as the *fly*, *twayblade* and *white helleborine* are in flower. As you climb up the slope towards the edge of the wood, a group of fungi with dark red-brown caps attract your attention. This is *spindle shank* that grows in dense tufts at the base of deciduous trees such as *beech* and *oak*. Unlike most other fungi which first appear in the autumn, this one can be found growing at any time from spring to early winter. Each stem, 4–9 centimetres high, is grooved and twisted along its whole length, is wide at the top and tapers to a thin stalk at the rooting-base. The caps are about 3–7 centimetres across and have white and brownish tinged gills. This fungus is not really edible because of its toughness.

As you climb the steps into the darker part of the wood you will see the remains of *bluebells*, their dry stems and brown seed pods belying the glory of these woods a month or so earlier. The woodland floor here is now covered with a carpet of oval-shaped and toothed, dark-green leaves; *dog's mercury*. This is one of the commonest of our woodland plants and grows to a height of 20–50 centimetres. Between the short leaf-stalks and the stems, spring strings of minute green flowers which soon adopt a drooping attitude. The flowers bloom from March until June, and by means of a creeping rootstock the plant forms extensive communities in woods. Dog's mercury is a member of the spurge family and shares the poisonous qualities of other species in that family. Livestock are well aware of this and often when dog's mercury is found in shade at the edge of a field, the lush grass around has been close-cropped, leaving the dark-green-lumps of the plant untouched.

Above the extensive wood is the summit of Newtimber Hill itself, now mainly chalk grassland, but from late Saxon times (800 AD) until the late Middle Ages (1500 AD) it was ploughed and cultivated by land-hungry peasants. As the population declined in the late Middle Ages so it was again returned to grassland only to be ploughed up again in the seventeenth century by farmers who sowed plants like *sainfoin* and *clover* for sheep fodder. Two hundred years later at the time of the Napoleonic Wars, when famine was rife, the land was turned over to cereals, only to revert back to grassland when prices fell as a result of the repeal of the Corn Laws. So the chalk grasslands and their associated flora came and went with the dynamic social changes in our history.

The most interesting part of this area is undoubtedly Saddle-scombe chalk pit, just to the south of Newtimber Hill, which is a small nature reserve managed by the Sussex Wildlife Trust. Here on its steep sides can be found small colonies of *juniper* and a variety of chalk grassland species including many orchids.

Juniper, an evergreen tree of the cypress family, is widespread throughout Britain, but local. It is found in many places on the chalk hills of Southern England but is extremely rare on the South Downs to the east of Brighton. It easily succumbs to fire and to scrub invasion and here at Saddlescombe, the site is carefully managed to ensure its survival. Juniper may be mistaken at first sight for gorse, but is easily distinguished by its bluish-green, spiky needles, which are grouped in threes along the twigs and covered with a silvery band of wax that restricts water loss. If crushed, these needles have a sharp and distinctive resinous scent. Junipers can live for 70 years or more, but seldom reach a height of more than about 5 metres. In spring, juniper bears clusters of yellow flowers, their wind-borne pollen easily received by the female flowers which are small green globes, hidden, like their male counterparts, amongst the foliage. By late autumn the female flowers have ripened to form round, dark blue berries. These are used to produce the volatile oil which is a prime ingredient of gin. The berries are used in many countries for other purposes; as a substitute for coffee and pepper, to make a health beer, and as an excellent survival food for man and animals because of their availability throughout the winter. Juniper berries were also once used medicinally, as a diuretic and as a carminative. A tincture of the branches was also used externally to treat many diseases. The wood of juniper was once burned to produce a charcoal used to make an extra-powerful gunpowder – savin powder.

One of the interesting plants to be found at the bottom of the chalk pit is the *red star-thistle*. This rare plant is easily recognised by its yellow spines surrounding the pale purple flower heads that bloom from July to September. Its lance-shaped leaves are greyish-green and woolly, and the whole plant is bushy in appearance, growing up to a metre in height. It is said that the sharp spines, with the seed-heads attached, get caught in the hooves of cattle, and this enables it to spread itself around the area. There are not many of these plants here in the chalk pit so you will need to look around for it. The Latin name for the thistle, *calcitrapa*, refers to the metal spikes which in Roman times were strewn across roads to injure the hooves of advancing enemy horses. On landing, these weapons were so designed that one spike was always pointing upwards, similar to the spines of the star thistle.

Before moving on to those lovely bare downs above Fulking and Edburton, do take time to explore the area around the chalk pit; see how many wild flowers you can discover and identify.

Walk across the deep combe ditch of nearby Devil's Dyke, so called from a Sussex legend that describes how the devil was so annoyed by the number of churches in Sussex that he attempted to cut this great dyke through the Downs in order to let the sea rush into the Weald to drown them all. Climb up to the top of the Dyke Hill above Devil's Dyke to the old Iron Age fort, and take in the glorious views all around you, trying at the same time to ignore the hotel and busy restaurant. The escarpment to the west shows a magnificent vista as it sweeps down in a series of curves and bluffs to the wide fields of the Weald. Little woods of *ash*, patches of *hawthorn* scrub, and areas of bare chalk, give character to this undulating escarpment, and the *whitebeam, gorse* and *tor grass* throw colour across its magnificence. Sadly, radio masts on Truleigh Hill, and stark pylons carrying high tension cables mar the view, but take the underhill lane beneath the escarpment and you will not be disappointed.

The lane winds its way through Fulking and past Edburton, with hedgerows of wild *dog roses* bordered by deep banks. Fulking was much loved by John Ruskin who visited it often when staying with friends in Brighton. The village is noted for its spring which bubbles out through a fountain by the roadside next to the Shepherd and Dog Inn. The spring was highly popular with shepherds who used it for washing their sheep. The fountain is a memorial to John Ruskin and has the following inscription on it:

> He sendeth springs
> Into the valley
> Which runs among
> The hills.
> Oh that man would
> Praise the Lord for
> His goodness.

The clear stream just below the fountain has dark green, flat fronds of *liverwort* growing about it. Liverworts, of which there are more than 200 species growing in Britain, are small primitive plants similar in structure to mosses, and grow chiefly in places where their delicate leaves will not dry up. Like *mosses*, liverworts reproduce by means of alternate sexual and asexual generations. In the sexual period, both male and female organs are produced giving rise to an egg, which results in a spore-producing capsule developing at the end of a stalk. When it ripens, it splits, so releasing the spores which develop into new plants. Some liverworts also reproduce vegetatively, where bud-like growths develop in special cups on the surface of the plant and these break away to grow into new plants.

Growing in places beside the edge of the lane you will see the lovely yellow flowers of the *large-flowered evening-primrose*. A robust and hairy plant, the large-flowered evening-primrose is about a metre tall, with leafy stems and large oblong leaves which form a rosette at the base. The large yellow flowers, made up of four petals, are about 5 centimetres in diameter. They appear from May to August and open up in the evening when their scent attracts night-flying moths which effect pollination. In 1949 it was discovered that the clear, pale yellow oil that came out of the seeds of this plant was rich in the fatty acids that are essential for the well-being of the human body. This evening-primrose oil is now used to treat patients suffering from nervous disorders such as multiple sclerosis.

At this point it is worth a short diversion to visit the headquarters and Countryside Centre of the Sussex Wildlife Trust at Woods Mill, some 3 kilometres up the A2037. Here, there is a nature trail laid out through a managed nature reserve taking in many habitats; woodland, marsh reedbed, lake, stream and hay meadow. The woodland is predominately *silver birch*, with *hazel* coppice and *oak*, in which grows a good range of wildflowers, including *bluebell*, *wood anemone*, *moschatel* and a colony of *common spotted-orchids*. *White admiral* butterflies breed in the wood as do birds like the *great spotted woodpecker*, *nuthatch*, and *tree creeper*. Summer migrants like the *blackcap* and *whitethroat* are well in evidence amongst the hedgerows and woodland edges, reed warblers breed in the reed beds, and the *kingfisher* nests beside the streams. There are good amphibian and dragonfly populations here as well.

An exhibition is housed in the headquarters building where information on all the Trust's nature reserves is available to the public. Do take the time to visit this reserve and to support the Sussex Wildlife Trust which does so much for nature conservation not only on the South Downs but in many other places as well. You will enjoy the atmosphere here where there are places for you to sit and listen to the sound of running water mingling with the songs of birds all around.

One is conscious of the South Downs not far away and on returning to the escarpment above Fulking you will see many more of the butterflies and wildflowers that you have now become familiar with.

It is worth recording that up until about 1800 the Downs here were once the principal haunt of the *great bustard*, a large bird that seldom took to the air. Sadly, these once relatively common birds were hunted to extinction for their meat by so-called 'sportsmen', who found them easy prey using greyhounds. Subsequently their gruff calls echoing across the treeless hills were seldom heard again in Sussex and the last sighting of them was in 1891 at Pett Levels. In recent years, attempts to reintroduce them in Wiltshire, using wild birds from Spain, have been unsuccessful.

South of the escarpment much of the land is cultivated, but you can still find little gems of natural grassland which are worth exploring, such as Anchor Bottom, just below Beeding Hill. In such places you may be lucky to come across one of our very rare and beautiful wildflowers, the *meadow clary*.

Related to the *wild clary*, mentioned in the previous chapter, the *meadow clary* grows up to a metre tall and has large striking, violet-blue flowers which grow in whorls of four or five around the stem. The flowering period is from June to September. The whole plant is moderately hairy and has soft and wrinkled, oblong leaves, which are jaggedly toothed. Because of the rarity of this plant the local farmer has been made aware of its existence so that it is given protection from grazing animals. I hope you will have the pleasure of finding it.

We have explored a large area in this chapter and it is time to move on westwards. From Truleigh Hill look out to Chanctonbury Ring above Washington which stands out prominently, with its small circle of dark trees. However, before exploring them it is worth taking the trouble to visit Shoreham beach and its wonderful flora; after all the beach itself is made up of flint pebbles, once encapsulated in the chalk of ancient downs which has been eroded away by earth movements and weather over tens of millions of years; well before man came on the scene.

For me Robert Bridges sums up this chapter of rolling hills and silent places:

> O bold majestic downs, smooth, fair and lonely;
> O still solitude, only matched in the skies:
> Perilous in steep places,
> Soft in the level races,
> Where sweeping in phantom silence the cloudland flies.

Chanctonbury Ring

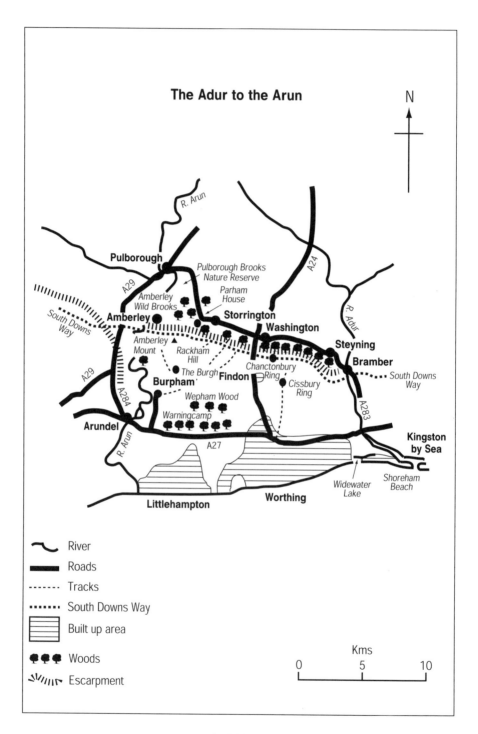

The Adur to the Arun

N

R. Arun

Pulborough

A29

Pulborough Brooks
Nature Reserve

Parham
House

Amberley
Wild Brooks

Amberley

South Downs
Way

Storrington

Washington

A24

R. Adur

Amberley
Mount

Rackham
Hill

Chanctonbury
Ring

Steyning

Bramber

South Downs
Way

The Burgh

Findon

Cissbury
Ring

Burpham

A29

A284

Wepham Wood

Warningcamp

A283

Arundel

R. Arun

A27

Kingston
by Sea

Widewater
Lake

Shoreham
Beach

Littlehampton

Worthing

~ River

Roads

----- Tracks

••••• South Downs Way

Built up area

🌳🌳🌳 Woods

Escarpment

Kms

0   5   10

118

6

THE ADUR TO THE ARUN

Amberley Castle

Surprisingly perhaps, we start this chapter at Shoreham Beach, a place of crashing waves and flint pebbles, formed over the past 60 million years or so as the chalk dome extending to France was split open by the upheaval of the great 'Alpine Storm', and was then weathered by torrential rain and frost. The extensive flint deposits within the chalk were gradually worn smooth by wave action into the pebbles that now cover much of the southern shoreline of England. Shoreham Beach itself is situated just to the west of the effluence of the river Adur into the English Channel. Earlier in the last century this was a remote place of extensive shingle and the haunt of many birds. Now, it is extensively developed as a residential area, but the beach itself, although devoid of the breeding birds it once knew, is a botanist's paradise.

Hundreds of years ago the Adur flowed into the Channel much further to the west, but just like the river Ouse at Newhaven, its mouth has been forced eastwards by the action of longshore drift. On the north foreshore of Shoreham beach, overlooking the untidy flotsam and jetsam of the estuary south of Kingston-by-Sea, grows a very rare plant indeed, the *childing pink*. Unusually, the site is enclosed by low railings and is well marked with a signpost describing the plant itself, and nearby is a convenient car park.

Sometimes I believe it is best to highlight a protected species in this way so as to prevent it being trodden upon, or otherwise despoiled. Flowering from June to October the childing pink has small, five-petalled pink flowers, clustered together in dense oval heads and each open one at a time. The stem is 15–40 centimetres high and the leaves are linear and hairless. It is not easy to spot hiding amongst the grasses of the shingle shore with more obvious flowers like the *Oxford ragwort, kidney vetch* and *dove's-foot crane's-bill.* Childing pink is a native of the Canary Islands, Morocco and Western Europe, and is known in Britain in only two places in West Sussex, including this site at Shoreham. It gets its name from its cluster of tiny flowers held in a papery sheath like 'childings' which is an old name for children.

The main beach looking out to the English Channel is only a few minutes' walk away from the site of the childing pink, and although a large number of private dwellings seem to bar access to the beach, there are quite a few access points from the road. The best time of the year to visit is in the last week of May, for at this time the shingle is ablaze with the glory of many and varied plants. The first one to catch your eye will undoubtedly be *sea kale*, one of the largest and most stunning of all maritime plants.

Sea kale dominates the beach here with its domed sprays of white flowers, 40–60 centimetres high, looking from a distance like outsize cauliflowers. It has bluish-green leaves which for centuries have been cut by local people, and when blanched, used as a substitute for asparagus. If cut when the leaves are young it is considered to be a great delicacy. Sadly the plant was often dug up to be grown in vegetable gardens and for this reason it became less common around our coasts. Now that this practice has been curtailed, the plant is once again on the increase and the main danger is the pressure of visitors to its shingle habitat. Interestingly, sea kale was well known to the Romans who gathered it from the wild and preserved it in barrels for use during long voyages.

The gardens of the seaside houses back down to the beach and contain many cultivated rock garden flowers of exotic beauty, and some, like a beautiful *erigeron* species which looks like a low-growing michaelmas daisy, have escaped onto the beach. It spreads across the beach and forms clumps of pink-petalled flowers with yellow centres. Another garden escape found abundantly here is the *red valerian* described in Chapter 3. It forms large patches of tall dark red flowers which stand out well against the backdrop of yellow-brown pebbles and the huge clusters of white seakale.

A tremendous variety of wild flowers thrive on this shingle

habitat, clearly finding ideal conditions. These plants are well suited to withstand the dynamic character of this seashore environment and are able to cope with the low nutrient content of the sparse and thin soil. So look for such species as the *yellow horned-poppy, curled dock, woody nightshade* and *ivy-leaved toadflax*, to name but a few.

The *starry clover* is the real gem of this important site. This clover has short stems and soft hairy leaves and the small white flowers which bloom in May and June are held together in rounded heads and each is surrounded by a crimson star-shaped calyx. When in fruit the flower heads with their white soft hairs, look even more like clusters of stars against the backdrop of beach pebbles. The whole plant is prostrate and forms large mats on the upper beach. Starry clover has been known on this site since early in the nineteenth century and is believed to have entered Shoreham Harbour in the ballast of ships returning from the Mediterranean. If you have not seen this plant before it may take you some time to spot it for the first time, but once you have 'got your eye in' you should easily find the places where it grows.

As you walk across this long, broad beach in your search for wild flowers, a movement just a hundred yards away catches your eye. A dog-fox (male *fox*), its beautiful red-brown coat radiant in the bright sun, strolls nonchalantly across your path towards the tall *red valerian* and the rear gardens of the seaside bungalows, no doubt looking for scraps of food thrown out by their inhabitants, or for dead sea birds and fish washed up by the sea. It passes so close that you can see its long pointed muzzle, erect ears and glinting eyes. About 60 centimetres long and some 32 centimetres high, the fox is the only wild relative of the dog in Britain. Once, many years ago it was only known as an animal of the countryside, but now it is increasingly common in towns, finding food easily from dustbins and scraps idly left strewn on pavements and in alleyways behind restaurants and bars, a trend becoming more common in our affluent modern society. In the countryside foxes make their dens, or earths as they are known, in a quiet place – in a hollow under a tree root, in a rabbit warren which they enlarge, or in an abandoned badger's set. A fox's earth is untidy, smells musky and often has food remains and excrement left around it; quite unlike that of the tidy and clean badger. Both the dog-fox and the vixen (female fox) spend solitary lives for most of the year, but they come together in winter to mate, and the three to six young are born in March or April. The dog-fox sometimes plays with his offspring, but it is the vixen that usually teaches them to hunt when they are about a month old. The young quickly learn the skills necessary to find food and to defend themselves, and within two or three months after being born, they are ready to leave the earth and lead their own lives. The dog-fox also leaves soon after the breeding season.

The fox, with its acute sense of smell, generally hunts at night, utilising the well-marked tracks it has made for itself. However, increasingly in modern times, it is losing its fear of man and is now often seen during the daytime. Foxes will eat just about anything they can catch, from *rabbits*, their staple diet, to *hedgehogs, rats* and birds; in winter when times are hard, they will eat beetles, worms, grubs and hibernating caterpillars. Here, on Shoreham beach, this remarkably tame fox is a real treat to see so close; he probably has his den in the scrub and wasteland around the nearby airport. You will almost certainly come across the animal again just about anywhere on the South Downs, and with patience, by sitting downwind of its den you may be lucky to see one emerge, and with careful stealth proceed to stalk and kill some unsuspecting rabbit.

A bird that was once quite common on lonely parts of the beach, is the *ringed plover*. Now it is not often seen here outside the winter visiting period of November to March when thousands of them migrate southwards to escape cold weather. About 19 centimetres long the ringed plover has a brown back with a black-collared white throat and breast. In flight you will easily spot its conspicuous white wing-bar which distinguishes it from a closely related species, the *little ringed plover* which nests in gravel pits, further to the west. The ringed plover's nest is just a slight hollow, scraped in the shingle above the high tide mark and generally there is no nest material, the four pear-shaped pale cream eggs, spotted with black and grey dots, being laid on the bare ground. Sometimes, however, the nest is lined with small pebbles or pieces of seashell. The eggs are extremely well camouflaged in their beach environment, and great care must be taken not to tread on them. The first eggs are laid in April, with repeat second and third clutches up to the end of August, and both parents take turns in incubation. After about 23 days the young chicks hatch out, and as soon as they are dry they are led away by the parents over the beach and well away from the nest site. The chicks are perfectly camouflaged, and if danger threatens, they vanish as if by magic, remaining quiet and still amongst the pebbles. Often, if dogs or humans get too near the nest or young, the parent bird will feign injury, by dragging a wing and shuffling across the ground. The potential predator will follow after the apparently injured bird, which, once satisfied that the danger is over, takes to the air. After about three weeks the young can fly and are then largely independent, feeding on worms, shellfish and insects; the parents are then able to start another brood.

Ringed plovers also nest occasionally on arable ground amongst cultivated crops on the tops of the downland south of the escarpment, but it is on the wide expanse of shingle beaches, particularly at Chichester and Pagham Harbours that they most often breed now. Here at Shoreham beach you may be lucky to see them and hear their melodious calls 'too-li, too-li', across the pebbles. Be aware, however, that these birds are most vulnerable to human disturbance during the breeding season, so if you come across them do not linger, and leave them to enjoy their peaceful surroundings and to raise their young successfully. Outside the breeding season, take pleasure in watching them as they run to and fro, their heads bobbing up and down as they search for food on the beach at low tide.

You may see many migrant and overwintering birds, not only on the beach itself but also around Widewater Lake just to the west. On calm winter days, scan the lake for *divers* and *grebes* and look for interesting waders, such as *Baird's sandpiper*, with its distinctive black legs and white belly; a rare bird which has been seen here more frequently in recent years. Sea watching can also be productive in spring and autumn with *pomarine skuas* appearing during onshore gales in early May.

I make no apologies for taking you away from the hills to see this truly beautiful place awash with the colour of early summer flowers. Before returning to the Downs sit on the shelving beach just above the water on a calm day and listen to the gentle lapping of small waves on the age-worn stones of the seashore. Ponder awhile and let the atmosphere and serenity of the place take away the troubles that lie deep in your thoughts. Here is a place where the primitive changelessness of the deep blue sea and the emptiness of the wide sky gives you peace and remoteness from the stress of everyday life; another place where you can recharge the soul and regain the strength to cope with your problems.

Cissbury Ring, our next destination, lies ten kilometres away, situated just to the east of Findon and was once the centre of Marnian culture in the south. These Marnians, the Iron Age Celtic invaders of 250 BC, built a huge fortification on the hilltop here, with a timber-reinforced rampart towering some 10 metres above the surrounding defensive ditch. About 2,000 years before, neolithic people had dug out long galleries in the chalk as they mined the area for precious flints. So Cissbury Ring has a long historical association with man and now it is an area of open downland managed by the National Trust. Although it is a popular spot with fine views, it has an interesting selection of chalk-loving plants and insects. In July and August look out for two beautiful plants belonging to the teasel family, the *field scabious* and the *small scabious*, both related to the *devil's-bit scabious* described in Chapter 4.

The *field scabious* reaches to a metre in height and has divided, hairy leaves and lilac-blue flowers which are bunched together forming tight heads at the end of long stems. The *small scabious* is a smaller version of the field scabious growing up to 60 centimetres in height and usually has somewhat paler flowers. The main distinguishing feature between the two is that the outer bracts of the flower head of field scabious are oval shaped, whereas those of the small scabious are long and narrow. Scabious is derived from the Latin *scabiosa*, itch, its juice being formerly used to cure scabies and other skin diseases.

An interesting moth can be found on Cissbury Ring and its surroundings; the *burnet moth*. There are seven distinct species of burnet moth in Britain and all have red spots on a greenish-black background with more than a hint of bronze colouring. This bronze-green tint varies with the direction of the light, as the colour is not pigment, but is caused by light waves interfering with the layers of scales covering the wings just as they do on a layer of oil on water. The moth you are most likely to see here in July and August is the *six-spot burnet moth* with its three pairs of bright red spots on each forewing. However, earlier in May and June, you may see the *five-spot burnet* moth. All the species are extremely poisonous, their bodies containing cyanide which is formed by the caterpillar from its food plants, stored up and passed on to the moths themselves. They are not harmful to humans unless eaten, but the poison acts as a powerful deterrent to predators such as birds, which are given ample warning by the red spots.

Burnet moths live in colonies and the female six-spot species lays her eggs in groups on the leaves of the *bird's-foot trefoil* plant. The caterpillars are speckled with yellow and black and feed on trefoils and vetches. They hibernate on these plants during autumn and winter and resume feeding in spring. In April they spin yellow boat-shaped cocoons, high up on grass stems in which they pupate. The moths hatch out in July when they can be seen in hundreds, their long thin wings, metallic colours and slow flight, easily distinguishing them from the numerous butterflies that flit amongst the grasses. It is uncertain how this group of moths got their name burnet, but one possibility is that the name comes from the association of the moth's red spots with burning.

Enjoy your visit to this place with its history and atmosphere. Leave enough time, however, to visit Chanctonbury Ring by taking the path northwards that intersects the South Downs Way. On reaching the intersection and strolling westwards up the slight slope, notice the wide grassy verges covered with wildflowers of all kinds, *wild basil*, *red bartsia*, *yellow rattle* and *knapweeds* to name but a few. Behind you the great ramparts of Cissbury Ring appear as shadows of some magnitude, and soon Chactonbury Ring itself comes into view, a circular defensive ditch with a few somewhat untidy and straggly-looking *beech* trees growing on and within it. These are the remnants of that majestic circle of attractive tall trees that once stood on this spot before the great storm of 1987; now most lie on the ground as grotesque skeletons of dead wood. The original beeches were planted within this prehistoric earthwork by a young Charles Goring of nearby Wiston in 1760. Then it was bare hillside, and it is said that he carried water up the steep hill for months to ensure that the young seedlings grew into the great trees they eventually became. Shortly before his death, some seventy years later, he wrote:

> How oft around thy Ring, sweet Hill,
> A Boy, I used to play,
> And form my plans to plant thy top
> On some auspicious day.
> . . . .
> And then an almost hopeless wish
> Would creep within my breast
> Oh! Could I live to see thy top
> In all its beauty dress'd.
> That time's arrived; I've had my wish,
> And lived to eighty-five;
> I'll thank my God who gave such grace
> As long as e'er I live.

The site on which he planted those trees was not only once an Iron Age fort, but was also used as a religious site and temple by the Romans in the third and fourth centuries. Similar temples, each about 7 metres square, surrounded by a verandah, can be found on hilltops where they were dedicated to local Celtic gods who were sometimes associated with the classical gods, Mercury, Apollo or Mars. Offerings were placed in these temples to the gods in the form of coins. On sites around Provence and west of the Rhine, such offerings were in the form of neolithic flint axes. No wonder local folklore has it, that anyone running seven times around Chanctonbury Ring on a dark night would be rewarded by the Devil himself, with a bowl of milk!

Public access to the centre of the ring is at present barred by fencing as young beech saplings are being planted to replace those destroyed in 1987. So in years to come this historic place will, it is to be hoped, regain its former magnificence. However, it is sad to see the sparseness of the land immediately surrounding this site itself. Here, few flowers can achieve their full potential and many are smothered by hard, long-lasting cow-pats. I last visited this place in August 1999 and then the place was littered with cattle dung which should normally exist in solid state for only about 20 days or so before being eagerly devoured and broken down by numerous beetles or bugs. On examination, the dung seemed solid and devoid of environmentally useful bugs, possibly as a result of the use of Ivermectin. This is a cheap and powerful anti-parasitic drug used by vets to kill worms in cattle and other livestock. It remains in the dung where it is effective in killing the bugs and allows the cow-pat to last for three months or more. Many birds, bats, and insects feed on fresh cow-dung and its contents. Interestingly, the National Trust banned the use of Ivermectin on around 100 farms in Cornwall five years ago. The main reason for this was to try to bring back the *chough*, a bird now extinct in Cornwall, but the Trust found that the ban had a far wider range of ecological benefit. As a result, the rarest of Britain's bats, the *greater horseshoe bat*, has increased in numbers on two Cornish sites and the equally scarce *hornet robber fly*, a hornet lookalike which feeds on fresh cow-dung, has also been found on two new sites. Perhaps the Ministry of Agriculture should examine carefully the effect of Ivermectin use on the ecology of our countryside.

As so often on this great northern edge of the South Downs, the views from Chanctonbury Ring and the nearby 'trig point' of Cross Dyke are simply magnificent. The Isle of Wight, nearly 60 kilometres away stands out as a bastion of land in a shimmering sea. In addition, the extensive wooded hills around Blackdown to the north rise majestically beyond Petworth as a bulwark of dark green and herald the fact that they are the highest part of Sussex just topping 279 metres. Just below lies the Lower Greensand, well marked by industrial sandpits, as it widens from a narrow strip

126

here to overlap the hills beyond Midhurst. The sandpits catch your eye with their clouded turquoise green ponds enclosed by shallow cliffs of sand, coloured with many shades of brown, yellow and orange. Far to the east, Wolstonbury Hill and the escarpment above Fulking and Poynings show up in a magnificent splendour of treeless downs.

It is late September and a flight of *house martins* passes by high above, swooping from side to side catching insects to sustain them on their long autumn migration southwards to Africa. A pair of *buzzards* glide lazily in ever-widening circles as they search for prey on the ground. Effortlessly gaining height on thermals set up by a warming sun, they seemingly ignore the harrying of a passing *carrion crow* that delights in diving upon them in turn.

The story of the *buzzard* in Sussex is interesting. Early in the nineteenth century this bird of prey was common in the county, but by the 1880s, persecution by gamekeepers virtually destroyed it as a breeding bird in the county. Gamekeepers, then, were intent on protecting the *rabbit*, the buzzard's main food source, because it was valued by man for its meat. The decline of the buzzard in Southern England was also aided by the spread of industrialisation and the reclamation of waste lands. After the First World War, the increase in the rabbit population and the reduction of pressure on gamekeepers who became more tolerant of wildlife, brought the bird back to Sussex. However, it was not until the 1950s that a small population was again breeding successfully. The introduction of myxomatosis into the rabbit population between 1950 and 1955, leading to its virtual extinction, was bad luck for the buzzard and only about four pairs were present in the county between 1950 and 1956. In recent years the rabbit population exploded once again as they became resistant to myxomatosis, and as a result the buzzard has now spread in greater numbers eastwards from its strongholds in Hampshire, Dorset, Devon and Cornwall. There are probably about 10 to 20 pairs now breeding around the South Downs in east Hampshire and Sussex.

The buzzard is easily recognisable in flight by its broad wings and large, rounded tail. Within its own territory, a buzzard has its favourite hunting perch on which it sits, watching for prey, well-camouflaged by its dark brown plumage. When it sees something, it moves its head from side to side, pinpointing the exact spot, then it descends in a gentle glide and finally pounces. The buzzard also hovers like a *kestrel*, and when its prey is sighted, it drops silently by partly closing its wings until, when just above the ground, it plunges rapidly onto the target. About ten unsuccessful attacks are made for each one that succeeds. After a kill the buzzard, unlike the kestrel, does not usually rise immediately, but can be seen on the ground, pecking at the prey caught in its talons. Large prey such as rabbits are taken to a perch for dismembering; smaller ones such as voles and mice are eaten whole.

The buzzard makes a substantial structure of branches as a nest and lines it with vegetable material, such as bracken, moss and green leaves. The nest is placed either on a cliff ledge, or more commonly in Hampshire and Sussex, in the fork of a tree from 6–20 metres off the ground. The two or three eggs are usually laid in April and the young hatch out after about five weeks in late May. This coincides with the emergence of many young rabbits, which are caught by both parents and fed to their fledglings. Let us hope that the buzzard population will increase in future years and contribute in an environmentally friendly way, to the control of the rabbit population which is now reaching an unacceptable level. You will see the buzzard frequently on your excursions over the Downs, especially in West Sussex, so enjoy watching these magnificent birds and their graceful flight pattern.

Time to move on, and into the magnificent beech wood just below the summit of Chanctonbury Ring. On your way down the steep, well-marked track, note the area of lightly-grazed grassland beyond the fence which is full of wild flowers and butterflies. Further down the slope an area of scrub has been cut back to prevent it from encroaching onto a particularly fine open area of pasture, which in September is coloured a delicate hue of blue from the thousands of *devil's-bit scabious* plants that grow on it. These flowers attract many beautiful butterflies like the *red admiral* and *comma*. As you descend into the wood a sense of peace and tranquillity is all around as you gaze up at the tall bare trunks of ancient beeches. In one place a small area has been planted with *western red cedar*, an evergreen tree introduced into Britain in 1854 from North America. It is the tree that was used by the native Indian to carve totem poles and from which he built canoes. The cedar has a red bark when young and its spray-like foliage resembles the flattish fronds of a fern. It is also a fast-growing, tall tree whose wood is now used for joinery and fencing. The dark and somewhat oppressive nature of these trees contrasts markedly with the open spaciousness of the rest of the wood, so you will feel the deep hushed stillness of these immature cedars, well sheltered as they are by the steep, heavily wooded slope above.

Not much grows under the dense canopy of this woodland except the occasional *herb-robert*, and of course *dog's mercury*, which thrives here to form a carpet of green over much of the ground. However, on the edges of the wood where there is greater light, many more flowering plants can be found. In one spot here, not so many years ago, a very rare orchid was found, *the lady orchid.*

The lady orchid has large, broad oval leaves which are dark green in colour and shiny and the flowers, which bloom in May and June, are pale pink or white with crimson spots on their broad labella. Close examination of these spots show that they are, in reality, bunches of tiny hairs. The lower labellum is divided into two square-shaped lobes, while the upper part is also divided into two lobes. The upper lobes, however, are long and narrow. The sepals of the flower form a dark reddish-brown hood over the top of the labellum. When viewed from the front, the large flower, some 2 centimetres from top to bottom, looks like a mid-nineteenth century lady, dressed as they were in those days, in a widespread gown and close bonnet; hence the name lady orchid. The flowers are held in dense spikes, up to 50 flowers on each spike, at the end of robust stems 30–60 centimetres in height. The whole plant is very conspicuous and is not easily overlooked when in flower. Sadly, it is very susceptible to the attention of rabbits which find the leaves particularly good to eat. The lady orchid is now restricted mainly to Kent with occasional records in Surrey and West Sussex. It seems to flourish in areas where filtered sunshine reaches the ground and is often associated with *dog's mercury, primrose* and *twayblade*. On your exploration of the woodland around Chanctonbury Ring look out for this beautiful and conspicuous orchid.

The South Downs Way with its undulating broad, chalk tracks and magnificent views over the north escarpment, supports on either side a rich wildlife habitat of grassland and deciduous woodland. On our way westwards to Rackham Hill, the *skylark* still sings its song high above you, and in contrast a solitary black *carrion crow*, glossy in the bright sun, greets you with a harsh croaking 'kaah, kaah, kaah'.

The *carrion crow*, about 45 centimetres long, is the great scavenger of the countryside, as its name suggests. It will feed on any carcass, small or large, and is particularly ready to eat eggs, young birds, frogs, earthworms and grubs of all sorts. Outside the breeding season it is sometimes seen in sizeable flocks, augmented by migrants from mainland Europe, particularly around rubbish dumps and along the seashores. However, as a general rule in the open countryside of fields and waste places, you are more likely to see it on its own, striding with slow gait across the ground looking for food. It makes its nest of sticks and mud, lined with grass, wool and hair, high up in the branch of a tree or tall bush. The four or five grey-green eggs blotched with shades of brown are laid from April to June. In recent years the population of this bird has increased markedly, probably due to the reduction in game-keeping pressures. The carrion crow is exceedingly intelligent and with its sharp eyesight and increasing population, it has spelt disaster to many ground nesting birds such as the *lapwing*. The carrion crow, despite its feeding habits, does not seem to have been affected by pesticides; indeed as its numbers increase so it is now seen commonly in urban areas where it has learnt to nest on building ledges and to take advantage of scraps of food left in gardens.

That beautiful, but deadly predator of small birds, the *magpie*, is a conspicuous feature on your walk. With its striking black and white plumage and long tail, it is easily distinguished from other members of the crow family, to which it belongs. The magpie is cunning and aggressive and has been increasing steadily in popula-tion since the decline in gamekeeping at the beginning of the last century. As with the *carrion crow*, but to an even greater extent, the magpie has spread into suburban areas where its population, certainly in the South of England, may exceed that for woodland or farmland. With its sharp eyes and intelligent behaviour, no fox, owl or cat can successfully stalk it and so it can be seen swaggering nonchalantly or in jerky undulating flight around our back gardens. I wonder how many people realise the immense damage a pair of these birds do to the relatively unprotected garden species such as the *blackbird*, *dunnock*, *wren*, and *robin*. I have often come across the broken eggshells and ragged and torn nests of these small birds and have heard the chattering 'chak-chak-chak-ckak' of a pair of magpies as they tear apart and devour young fledglings still in their nest. The magpie's nest is well protected and is usually built in a thick clump of thorn bushes. It consists of a foundation of dead bramble or thorn-sticks firmly plastered together with mud, on which is constructed an inner cup of mud, lined with fine roots. A dome of thorny twigs with an entrance hole completes this large and formidable fortress 'home', which is quite impene-trable to any potential enemy and gives good reason why this bird is one of our most successful species.

One of the butterflies that you may see over the Downs here, especially in July, is the *marbled white*. Despite its name and colour, this is one of the brown family of butterflies like the *meadow brown.* The marbled white is a black butterfly with white spots, rather than a white butterfly with black spots. Two centuries ago it was known by country people as the 'half-mourner', because women then, wore black and white dresses during the period of 'half-mourning' which followed full mourning for a dead relative. Its name then changed to *marmoress*, meaning marble-like, and finally, about a hundred years ago, it became known as the marbled white. This butterfly likes ungrazed, or only lightly grazed, grassland slopes, where the adults feed on the nectar of such plants as *scabious, knapweed* and *marjoram*. For this reason they are found more frequently on the eastern Downs than on the more wooded slopes further west. They live in colonies, which at one time were far more widespread, before farming became intensive, and many of our flora-rich grasslands were converted into grass made suitable only for sheep grazing by the use of artificial fertilisers. In July, the female, which is slightly larger than the male, scatters her eggs amongst tall grasses while in flight. The caterpillars hatch out in about 18 days and almost immediately begin a long hibernation until the following spring, when they commence feeding on grasses. In June they form a chrysalis and pupate for about three weeks, the adult butterflies then emerging to repeat the cycle. So look out for marbled white butterflies on your walks as they show off their bright contrasting colours with wings spread open on colourful wildflowers. Smaller colonies of them can also be found on the verges of West Sussex roads and in woodland rides, probably remnants of the large populations that once existed.

Wherever you come across the marbled white you will almost certainly see its relative, the *meadow brown*. This is the commonest of our butterflies and is most often seen between May and September flying over grassland and roadside verges of the Downs, the Weald and the coastal plain. The female meadow brown, with its colouring of orange and shades of brown, is larger and more colourful than the male which is much darker overall. Both sexes have a prominent false eye on each forewing, although in the male it is less prominent. These 'eye-spots' are said to confuse its main predator, the bird. A bird will tend to attack the eyes when it swoops down upon the butterfly, as it may surmise that once the eyes are destroyed, the creature would then be at its mercy. All the brown species have false 'eye-spots' and when a bird attacks, many escape with just a damaged wing as the bird pecks at the prominent false 'eyes'. Thus over aeons of time those with the best 'eye-spots' survive and pass on their genes to their successors. Even on a dull day, when most other species are dormant and hiding away in the long grass, the meadow brown will rise up and fly around in front of you. Like most butterflies, the meadow brown has a short life span of just three or four weeks and there is just one generation each year.

As we approach Rackham Hill along the ridgeway, the great stone mansion of Parham House, built in 1559, with its fine furniture, textiles and paintings, is set in wonderful parkland just below you. To the west, the great wilderness of wetlands, Amberley Wild Brooks, catches your eye and interest.

Rackham Hill itself is bounded by a fine prehistoric terrace-way which traverses up the face of the escarpment and is lined on each side near the summit by banks over two metres high. These were said to have been built to prevent cattle from straying into adjoining cornfields when being driven from one grazing area to another. In the spring these banks are covered with yellow *cowslips*. Later in June you will find here, plants such as the *common rock-rose* and *fairy flax*.

The pale yellow flowers of the *common rock-rose*, 2–3 centimetres across, are abundant on chalky downland banks from June to September. This member of the *Cistaceae* family is often confused with buttercups, with which it has no relationship. The five petals of each flower are not glossy as in the buttercup and are often crumpled and have the appearance of being soft in texture. The whole plant is shrubby and its branches trail along the ground among grass and low vegetation. The small shiny, oblong leaves are a downy white underneath, and they grow in pairs on the stem. It is thought that the common rock-rose could be a survivor from the time of the intense cold of the late glacial period of 14,000 years ago. It still grows on cliffs and rocks in the Scottish Highlands up to 640 metres above sea level, quite a different habitat from the soft rolling hills of the South Downs.

Fairy flax is a slender plant with thin, wiry stems and opposite pairs of small oblong leaves. About 10 centimetres in height, it is very common on chalk grasslands and has small white flowers which appear from May to September, which are succeeded by round seed capsules. Fairy flax is also known as *purging flax* because it was once used as a laxative, the bruised stems being boiled in water and white wine, with peppermint sometimes added to hide the bitter taste.

On a still summer evening the atmosphere on Rackham Hill and Amberley Mount is one of great splendour as the sun sinks over distant dark hills, reflecting perhaps a glint here and there off the water-meadows of Amberley Wild Brooks below. A gentle breeze drifts across these open places and winnows the fragrance from grassland flowers like *marjoram* and *wild thyme*. In winter the atmosphere is more melancholy when these gentle hills seem devoid of life. There are no colourful flowers in the grasses and to the south the fields are left bare brown by the plough, and only large flints remain to break up their monotony with flecks of white. From Amberley Mount, the river Arun is seen in silvery snatches as it flows swiftly but smoothly and untroubled, south to Arundel. In early spring the harsh noise of a tractor far below breaks through the awakening stillness of nature. White specks gather behind it looking like blossom falling from a tree; gulls ranging far inland from the sea are foraging for worms turned up by the plough. Stark, leafless lines of hedgerows await the warmth of spring as the earth tilts on its axis and the sun climbs remorse-lessly along its azimuth. Soon the leaves will break out, the hedge-rows will turn green, and white starry flowers of *blackthorn* and *hawthorn* will add flecks of brilliance to the scene.

The lovely village of Amberley lies just below the Downs here, and beyond to the north, where the river really widens into an alluvial plain, lies the marshy world of Amberley Wild Brooks. The 'wild' conjures up in one's mind a lonely, natural world of untamed water-meadows, and that's just what it is, even though the word wild is here derived from Weald. This is a place where history and time stand still, a place to linger and watch wildlife of many kinds amongst the ditches, streams and half-hidden pools which merge with reed-filled pastures and patches of sedge. The character of this area changes with the seasons and holds a fascina-tion for the naturalist at all times of the year.

The Wild Brooks cover an area of some 1000 acres, some 200 of which are owned and managed by the Sussex Wildlife Trust. The whole region is a wide, flat, alluvial flood-plain of the river Arun. The central part was originally a raised peat bog and has acid waters supporting somewhat different plants to the calcium-rich waters elsewhere. Thus, the Wild Brooks has one of the most bota-nically diverse systems in Britain.

In winter the marshes are wide open to cold north winds which burn their seeds and grasses into many shades of brown. Winter, with its flooding, also brings waders and wildfowl of all kinds with significant numbers of *teal, shoveler duck* and *Bewick's swan*. The latter has one of its most important sites here, where it grazes on various areas of agricultural land. The Bewick's swan is considerably smaller than the *mute swan* which is so common on our lakes, rivers and streams and it has a rounded area of yellow on its bill quite different from the orange of the mute swan's bill. The only other swan that the Bewick could be confused with, is the *whooper swan*, but here again the Bewick is much smaller and its neck is shorter; also the yellow on the Bewick's bill forms a point and is less extensive than the yellow on that of the whooper swan. Bewick's swans do not nest in Britain, their breeding haunts being in the wilds of Russia and Siberia. They migrate to western Europe arriving in Sussex late in October and departing in March and April. What a picturesque sight they make as large herds of white birds ride on the grey waters of the wetlands, occasionally uttering goose-like gabbles. In spring, at least 57 species of bird breed on the Wild Brooks including the *lapwing, redshank, snipe* and *yellow wagtail.*

The *lapwing* is one of our nationally declining breeding species; the reason for this is not known for sure, but a combination of factors is probably the cause. Its favourite nesting habitat in Southern England is mixed farmland where it shows a preference for spring-sown cereal crops with adjacent grass fields for feeding. It also nests on downland fields, particularly in the area south and south-east of Amberley. In March and April it makes a small scrape in the ground, lines it with grass and then lays four, well camouflaged eggs, coloured olive green and blotched all over with blackish-brown. These eggs are particularly vulnerable to predation by *foxes, badgers, crows* and *wild mink,* the latter being a comparatively recent pest which has spread from fur farms and become naturalised. This predation, together with the lack of suitable undisturbed habitat, are the most likely causes of this most attractive bird's decline. Let us hope that here on Amberley Wild Brooks the population will stabilise and perhaps increase, and visitors will forever be able to see its acrobatic courtship display flight in the spring and hear its repeated and plaintive call, 'pee-wi, pee-wi'. The lapwing is called peewit by country people and is still a common sight in autumn and winter when large flocks migrate from northern pastures to feed on invertebrates and vegetable matter in the river valleys of the South Downs.

The *redshank* is both a resident and a summer visitor and is common in most parts of Britain where there are lush water meadows, salt marshes, or rough pastures at the edge of moors. In such places, it arrives in the early spring with its noisy, yet alluring, whistling cries. The redshank can be distinguished when at rest by its orange-red legs, and in flight by its conspicuous white rump and broad white stripe at the rear edges of the wings. Both sexes are very much alike. The spring courtship display of the redshank is one of beautiful balance and symmetry. After a short chase, when the male, with tail fanned out, moves crab-wise across the ground, the female pauses in her tracks. The male opens its wings almost vertically above its back, displaying the white undersides to the female. The male then begins to flutter its wings, singing repeatedly 'tleu-tleu-tleu' and approaches the female with exaggerated high steps and head bowed. Eventually mating occurs and the first egg is laid 24–48 hours later. Early courtship display is evident when the male rotates his breast in a tussock of grass making a nest scrape. Several such scrapes are made and the female then selects the scrape in which she will lay, and then lines it with grass stems and moss. Interestingly, the onset of mating is correlated with differences in spring temperature and any sudden rise in temperature above about 8°C brings about egg-laying some five days later. The four attractive, buff-coloured eggs, each with dark, bold blotches are laid in April. Sadly, many nests are destroyed by grazing cattle or sheep, and unlike the lapwing, the redshank will seldom attempt to attack and drive them away. Other predators like the *crow* and *fox* also take their toll, but replacement eggs are laid until the end of June. Both adult redshanks take turns in incubation and the young hatch out after about 22 days. The chicks all hatch out within a day and leave the nest as soon as they can run, a matter of hours in most cases. Both adults and young feed on a variety of small insects, spiders, worms etc., and also on vegetable matter such as seeds and some buds. Within a few months, as soon as the young are strong on the wing, adult birds and juveniles make their way to coastal estuaries where most spend the autumn and winter. However, some migrate across the Channel to the warmer parts of Europe, particularly when severe weather is encountered. A good place to see wintering redshanks is the mudflats of Pagham and Chichester Harbours when sometimes there are as many as 3,000 individuals.

The Arun valley, and particularly the Wild Brooks, is the main breeding stronghold in Sussex for the *snipe*. This bird is most often seen when it gets up suddenly out of marshy vegetation from only a few feet away. Its flight is a characteristic zigzag pattern and as it rises in a fast, but erratic manner, it utters a short sharp, rasping note 'schaap'. Seen at close quarters, the snipe has a long straight bill, and its plumage is patterned with black, red-brown, cream, and golden buff arranged in stripes down its back and on its head. It is about 31 centimetres long, including its long straight bill. It is normally a secretive bird, but in the early spring when it returns from its wintering grounds near the coast, it claims its breeding territory by a peculiar display flight. It rises high in the sky over the marsh and suddenly dives, the wind vibrating through its outspread tail feathers producing a characteristic drumming noise. This vibrating, bleating noise serves as its song as it repeatedly rises and dives, until it has circled its territory. Like the redshank, the snipe makes its nest by hollowing out a tussock of grass or rush, and then lines it with a few dry grass stems or other herbage. The four beautifully camouflaged eggs are laid in April and May. Both the male and female birds, which are almost identical, share the incubation and the raising of the chicks.

Like the lapwing, the number of breeding snipe has declined and again it is the loss of habitat that has been the main cause. Snipe favour poorly drained pastures that are subject to winter flooding to provide them with a plentiful supply of worms and insects. Sadly, modern farming methods require a lowering of the water-table on low-lying pasture land to give earlier and more intensive grazing for livestock. This, of course, is the exact opposite of what the snipe and other waders require, and so the quality of their breeding habitat is reduced. Luckily we have such nature reserves as Amberley Wild Brooks where the management is in the hands of trained naturalists to ensure that the needs of the birds come first.

The *yellow wagtail* is a true migrant that arrives in flocks in April from its wintering area south of the Sahara. About 15 centimetres long, this attractive bird has a yellow and olive head, with a bright yellow eye-stripe. Its throat and underparts too are of the same bright yellow while its cheeks and upper parts are more olive green. The yellow wagtail is a bird of the open flat marshland, where, with its delicate colouring, it looks lovely as it flits and takes little runs, chasing flies and insects. The flocks soon break up as the birds pair off and build their nests of grass and rootlets, lined with hair or feathers under a tussock of grass. Four to six greyish white eggs, thickly speckled with buff-brown, are laid in May or June with a second brood in July. After the breeding season is over, the birds keep together in family groups but as autumn approaches they begin communal roosting and soon they form flocks of ten or more. They become increasingly restless until by the end of September most have departed for their southerly migration. I hope you will catch sight of these birds with their exciting musical calls 'tsweep tsweep' across the marsh, or perhaps see two rival males fight each other, tumbling to earth locked together in combat. Be careful not to disturb them, particularly in the breeding season as they, like so many others, are a declining

136

species, which are dependent on the seclusion afforded by this reserve and by Pulborough Brooks just to the north.

Of necessity I have described only a few of the many bird species to be seen on the Wild Brooks. With patience and by careful, silent observation you will see many more, and whatever the season you will enjoy the haunting atmosphere of this lonely wilderness of reed, marsh and water.

Early summer is a good time to see the many wild flowers of the Reserve. As with the birds, it is difficult to choose from the multitude of species but my favourites are the *frogbit, flowering rush, greater bladderwort* and *purple-loosestrife.*

Frogbit is a pretty aquatic plant with white and yellow three-petalled flowers which could be confused with the *water crowfoot,* but that more common plant has four petals. Also frogbit flowers in July and August just after the crowfoot has completed its display. Frogbit floats upon the surface of still, calcium-rich waters of the marshland ditches. It sends out long horizontal runners which produce bunches of roots which descend and penetrate deeply into the mud; from some of these roots small bulbs are produced, from which new plants arise. The runners also produce groups of kidney-shaped leaves, which help to highlight the flowers that grow on stout stalks above them.

The flowering rush is a handsome plant, 1–2 metres high, that grows amongst vegetation on the edge of ditches. It is easily recognised when in bloom on account of its clusters of large rosy flowers, 3 centimetres across, that are held at the top of the long stems. Before it flowers in July and August, it is easily overlooked because its long slender sword-shaped leaves merge inconspicuously with other aquatic vegetation. Its Latin name, *butomus,* is derived from the Greek words *bous,* an ox, and *tomos,* cutting, implying that oxen are sometimes cut by the leaves of the plant when they are drinking.

A plant that is becoming increasingly uncommon, is the *greater bladderwort.* Its decline is mainly due to the loss of habitat with improved drainage and by the use of agricultural chemicals and fertilisers, which pollute our rivers and streams. The greater bladderwort is an insect-eating water plant which has no roots but floats in the calcareous water of the many ditches on the Wild Brooks. In July and August its rich yellow flowers may be seen rising above the surface attached to a leafless stalk some 10 to 20 centimetres tall. The finely divided leaves lie submerged beneath the surface and attached to them are small pitchers or bladders. These bladders have elastic door-like valves which open inwards allowing small aquatic animals to enter, and then close behind, so trapping them inside. The plant then digests the animals as they decompose. The leaves with attached bladders of one plant can cover an area of over 100 square centimetres and so have a good catchment area to acquire their food. The bladderwort rarely sets seeds and usually spreads by detached pieces floating away to form new individual plants. In winter, the buds of the plant sink to the bottom until spring arrives when growth begins again.

A beautiful sight growing amongst the vegetation along the miles of ditches and streams and in the marshlands, are the tall, handsome spikes of *purple-loosestrife* with their numerous starry, reddish-purple flowers. The angled stem can attain a height of over a metre and bears lance-shaped hairy leaves arranged in pairs or in whorls of three or four. It blooms from June to August and is pollinated by various long-tongued insects including many species of butterfly.

Amberley Wild Brooks support some 38 species of freshwater molluscs, some of which are national rarities. There are also 16 breeding species of dragonfly, many of which you will see as you walk the public footpath known as the Wey-South Path, that runs northwards from the village through the Wild Brooks.

Dragonflies are amongst the fastest-flying and oldest insects in the world, with flight speeds up to 100 kilometres per hour. Fossilised remains show that great dragonflies with wingspans of 60 centimetres existed in the Carboniferous epoch of some 240 million years ago. As we watch these lovely flying insects with their transparent wings darting around the ditches it is worth reflecting that our earliest ancestors have only been in existence for one million years or so. Dragonflies are efficient hunters with their enormous eyes which cover the top and sides of the head. They feed on flying insects up to the size of small butterflies, which they catch and eat in flight, although sometimes they perch on a fence post to eat the larger ones. Often, you will see two dragonflies flying together in a mating embrace. The female lays her eggs in the water which then hatch into nymphs or larvae. These nymphs live in the water for up to two years, feeding off creatures from tiny insects to large tadpoles, depending on the size of the nymph itself. When mature, the nymphs leave the water by climbing up plant stems in early summer. They then shed their skins and emerge as winged adult dragonflies which only live for about a month.

Two dragonfly species to look out for are the *broad-bodied libellula* and the *common aeshna*. The former has a wingspan of 76 millimetres, and a length of 45 millimetres; its broad abdomen is

138

mostly blue in the male and is brown with yellow edges in the female. The larger, common aeshna has a wingspan of 95 millimetres and a length of 76 millimetres. The male has blue spots on its abdomen, whereas the female is less striking with dull green spots on its body. So often in the natural world, the female is duller in colour than the male in order to be less conspicuous to predators, which in the case of dragonflies, are birds like the *wagtails* and the *hobby*.

Various species of *damselfly* will also be found on the wetlands here. These smaller relatives of the dragonfly can be recognised by the way they rest with their wings above their bodies while the dragonfly rests with its wings straight out beside the body.

As you walk the quiet path with water either side of you, a sudden 'plop' and a ripple on the surface catches your attention. This was possibly a *water vole* diving from the bankside into the dyke and scudding away under the surface and out of sight. This species is the largest of the British voles measuring about 20 centimetres long. It is sometimes incorrectly called the *water rat*, a confusion with the *brown rat* which is also a good swimmer. However, the vole is easily distinguished from the rat by having a much shorter and furry tail, shorter ears and a rounder less pointed face. The vole also swims submerged, whereas the rat tends to keep on the surface. The diet of the water vole is mainly vegetable and one sign of its location is a patch on the bank where the plants have been bitten off. Apart from the breeding season in spring, the sexes remain apart and both hold separate territories each, having a stretch of bank, the male's being larger than the female's. The nest is made of grass or rushes at the end of a burrow in the bank, the entrance of which is sometimes just below the surface of the water. Two to four litters of usually less than six young, are produced each year, the first being in May. The water vole has many enemies in nature such as the *weasel*, *fox*, *pike*, *adder*, birds of prey, and particularly the non-native American mink (*wild mink*) that has escaped and spread from mink farms. However, the main reason for its rapid decline in recent years is man himself. We have allowed many of our watercourses to dry out due to excessive underground extraction of water by the water companies. Many of the river bank habitats of the water vole have been 'tidied up' by cutting the bankside vegetation, and this vegetation has also been destroyed by concrete embankments built to prevent erosion. Pollution of our water courses, too, has led to the vole's demise in so many places. Luckily, the problem has been recognised and much is being done by our conservation bodies to bring back and encourage the return of this lovely mammal.

It is time to leave the unspoilt wilderness of reeds and rush-filled pastures and return to the Downs above Amberley. As you walk southwards along the Wey-South Path, the sheer walls of a castle rise dramatically from the edge of the marsh. This is Amberley Castle, which was built as a manor house for the bishops of Chichester early in the thirteenth century. Happily no wars or sieges came to this place and now it is a pleasant and attractive hotel. On the walls of the castle in May and June grow *wallflowers* which are not true native British plants but were introduced in the Middle Ages from southern Europe as garden plants. The wall-flower is now firmly established on old ruins throughout the country and is also found growing on chalk cliffs near Beachy Head and at Lewes. The wild form of this species, here at Amberley Castle, has very fragrant, four petalled, yellow or brownish-orange flowers. It is said that wallflowers were planted on the walls of manor houses and castles so that their perfume could waft through the open windows and give pleasure to the occupants.

Retracing our steps up the steep slope of Rackham Hill topped at this time with billowing white cumulus clouds growing taller by the second, you can look back at the small wilderness of wildlife, now hardly touched by the ravages of man; long may these Wild Brooks remain so.

Autumn is here and our destination is the little village of Burpham, just to the east of Arundel, which is reached in a variety of ways by different ancient paths. The view to the south is spectacular. Across a landscape of cultivated fields and hills lies the sea, a silver sheet of reflected sunlight that stretches beyond the flat coastal plain and meets the sky in a wide unbroken horizon. Take in the rich autumn colours of browns and yellows and the curious, somewhat mystifying half-light that brings out the green intensity of the Arun valley. Across the valley, sheets of heavy rain, falling dark grey beneath the ever growing clouds, give a slight sense of menace to the scene. A small buffish-brown bird perched on the branch of a hawthorn tree just ahead of you, catches your eye. It seems unconcerned by the approaching storm, as it flits from branch to ground and back again. You notice its prominent buff eye-stripe and white patches at the base of its tail and recognise it as a female *whinchat*. This species is of similar appearance and behaviour to the *stonechat* described earlier, but is easily distinguished by its eye-stripe and less plump form.

The *whinchat* is widely but locally distributed throughout Britain during the spring and summer, and was once a consistent breeding species on the South Downs, where it favoured rough ground with gorse and scrub, particularly on south facing slopes. Sadly, because of habitat destruction it has long since ceased to nest on a regular basis. The whinchat is a migrant from Africa which arrives in April and May and departs again in August and September. Clearly, this one is feeding up on insects, such as flies and spiders, and awaiting favourable winds to help it on its migration south.

As you walk along the path, you will see on either side large fields where cereal crops have long been harvested and are now covered with wild flowers, such as that ubiquitous weed, the *prickly sow-thistle* with its dark, glossy spiny leaves and yellow, dandelion-like flower heads. Other common plants here are *common toadflax, field pansy* and *scarlet pimpernel*. Weed seeds are plentiful and flocks of *linnets*, with their characteristic undulating flight path and twittering calls, make the most of this plentiful source of food. In particular, linnets like to feast on thistle heads.

The *common toadflax*, as its name suggests, is widespread on dry banks and field borders beside downland tracks. Up to 60 centimetres tall, it stands out from afar with its bright yellow long-spurred flowers which have a snapdragon appearance. *Yellow toadflax*, as it is sometimes called, has stiff stems with numerous narrow pointed leaves. The flowers which bloom from June to as late as November have a touch of orange on the 'palate' which attracts long-tongued bees. These have to prise open the lips of the flower to obtain the nectar contained at the end of the spurs, and this requires some effort. Occasionally you will find flowers with holes bitten into the bottom of the spur, where a bee has discovered that it can obtain the nectar more easily by biting through from the outside. Another toadflax that you may also find here in waste places beside the cultivated fields, is the *small toadflax*. Much smaller than the common toadflax, this species is stickily-hairy all over its much branched stems, and has narrow, blunt, alternate leaves and purplish 'toadflax' flowers. It, too, blooms from May to late autumn.

The *field pansy* was once found just about anywhere amongst growing crops on chalky fields. With the widespread use of herbicides in the 1950s and 1960s this, like so many other common species, declined in numbers. Luckily farmers are now more aware and concerned about conservation matters, and frequently leave a space around field boundaries which are not sprayed. This allows species like the field pansy to survive and indeed flourish. Here, north of Burpham, it is a very common plant, with its small flowers displaying a great variety of colours from blue-violet to creamy-white and yellow, or a combination of all these; it has lance-shaped leaves and is in flower from April to November.

The bright red, five-petalled flowers of the *scarlet pimpernel* are as conspicuous as they are well known. It lies along the ground and has a square stem with oval stalkless leaves which are usually borne in pairs along the stem. It has long enjoyed a reputation amongst country-dwellers as a cheap barometer because of its habit of closing its petals on the approach of rain. However, this is only reliable in the morning after about 8 a.m. as invariably the flower closes up soon after 2 p.m., come rain or shine. It is also known as old man's weathervane and poor man's weatherglass.

Take your time as you wander south amongst these gentle hills. The trackways takes you through tranquil areas of high hedges where *wild clematis*, with its white feathery seed-heads, climbs up and over the tops of *hawthorn* and *dog rose*; no wonder it is called *old man's beard*. Its snow-white seeds last until late in the winter, and when all the trees and hedges around are bare, its presence gives colour and pleasure to the walker; not surprisingly it is also known as *traveller's-joy*.

Almost by chance you stumble upon what appears to be an old dew-pond, dried up with clumps of *gorse*, *dogwood* and *dog rose*. Around its edge grow downland grasses, close-cropped by *rabbits*. This is an interesting habitat for wildflowers which is worth a closer examination, as the thunderstorm, that had threatened for some time, expends its fury and passes by well to the west. On your map you notice the area around is called 'The Burgh', and nearby is a tumulus, its circular, raised mound of earth, clearly visible beside the track. Not far away earthworks of early man are in evidence; no doubt this was a place of some importance a long time ago. Indeed, a Bronze Age urn, pieces of bone, and much Roman pottery has been recovered from this site.

A study of the area within the rim of the dew-pond, reveals a multitude of interesting plants. Two orchids are still in flower, the *frog orchid*, somewhat surprisingly this late on in the year, and the *autumn lady's-tresses*. Other species to be seen are *carline thistle*, *felwort*, *small scabious*, and *harebell*. Of all these, the one not yet described is the delightful autumn lady's-tresses.

As I mentioned in Chapter 4, the autumn lady's-tresses is often found on the garden lawns of estates built upon what was once natural downland, but here on this ancient site, it flourishes not as an orchid of suburbia, but as a true native of our countryside.

The autumn lady's-tresses is a charming plant of fairly dry, chalk soils where it flowers in August and September. Its white flowers grow in a near-perfect spiral around the stem which is about 8–15 centimetres tall and is covered with short white hairs. The flowering spike appears to have no leaves at its base, but if you look carefully, you will find the withered remains of this year's leaves. Close by the stem base, amongst the grass, you will also discover a hollow roll of three or four very fresh, developing oval-oblong leaves which are in fact those belonging to next year's flower spike. The numbers of this orchid that may appear, fluctuate considerably from year to year; sometimes a known locality will be covered in thousands to be followed the next year by hardly any at all. The factors that cause these fluctuations are not understood, but weather and mycorrhizal activity are certainly involved. My own observations suggest that a wet March or April leads to a profusion of flowers in the summer. This also seems true for other orchids that have a tendency to fluctuate from year to year such as the *early spider orchid*.

Time to leave this idyllic spot where within its sheltered hollow there is so much to see and examine. Many tracks lead south to Burpham and it is difficult to choose which one to take. The summits of surrounding hills are bare and gently rounded. In places, but well-spaced, are little dark woods and small copses of lighter green. The vegetation beside the paths is lush even at this time in autumn, and here and there it is splashed with the colours of many flowers. Suddenly a covey of partridges explodes out of the grass just ahead; two, three, five, and then six, quickly take to the air flying with fast wing-beats, low over the ground, before gliding to alight some distance away. They allow one little time to determine whether they are the native *grey partridge*, or the *red-legged partridge* which was introduced in 1770 from south-western Europe. Both species can be seen in the area and are fairly widespread on the South Downs, particularly on agricultural land where their staple diet of weed seeds, grain, clover leaves, and insects are abundant.

When seen at short range these two partridges can be easily identified, the grey partridge having a pale chestnut head, grey bill and neck, whereas the red-legged species has a grey crown, white cheeks and throat, and a red bill and legs. Red-legged partridges are seen in less compact flocks, and are more nervous than the grey partridges; their movements on the ground tend to be quicker too. When flushed from cover they utter a 'kuk-kuk-kuk' whereas the greys have a sharper cackle, 'kri-kri-kri'. Just to add some confusion to the partridge story, in 1970 more than 2000 captive bred partridges of two different varieties known as 'ogridges' and 'chukars' were released into the wild on the South Downs just south of Washington. Releases continued until 1992 when the licence under which the practice was allowed, ceased. These introduced birds are similar in looks to the red-legged partridge and since they have undoubtedly bred with, and diluted the stock of 'pure' wild partridges, it is difficult to determine the present situation.

Both species of wild partridge make their nests as scrapes on the ground in hedgerow vegetation or amongst corn or meadow grasses. Ten to twelve pale olive eggs (speckled with fine dark spots in the case of the red-legged partridge) are laid from April to June. The female alone incubates the eggs, but after they hatch in 24 days or so, the cock bird helps to guard the tiny, fluffy red-brown, buff-spotted chicks, escorting them out into the fields to seek insects which form so much of their early fare.

The grey partridge has declined greatly in numbers since 1950 due to post-war changes in agricultural practice. The increased use of herbicides in cereal crops reduced the insect food available for the chicks, and the removal of hedgerows and much of the field boundaries in our countryside diminished the places where partridges could safely nest. Unlike the red-legged species, the grey is not artificially stocked for shooting purposes, and there is little reason to suppose its decline will not continue, unless, of course, there is a definite will to give it the conditions and habitat that it requires.

Proceed slowly southwards, and linger in places to absorb the atmosphere of these lonely Downs, beautiful in their open spaces. This is the last time we shall see them so, for beyond the river Arun they are more thickly wooded, still beautiful, but without the sense of space and remoteness further east. For me, that great poet John Masefield sums up this part of the South Downs in his lines:

> Something passes me and cries as it passes
> On the chalk Downland bare.

As you enter Burpham you will not fail to notice showy spikes of yellow flowers growing beside the hedgerows. This is the *dark mullein* which is an altogether smaller and less robust species than the *orange mullein* and the *great mullein* mentioned in Chapter 4. Often a metre tall, this species has an angled stem and its leaves are less hairy and do not have the velvety texture of the *great mullein*, and its yellow flowers which appear from June to October have purple stamens at their centre which give it the name dark mullein. It is fairly common on banks and roadsides on lime-rich soils but for some reason is found more frequently on the western part of the South Downs, and is an uncommon plant in East Sussex.

Burpham has that quality of timelessness about it, an enchanting village that has been left alone for centuries. Steeped in history, it lies at the end of a narrow road that runs north from the A27. It was built around an ancient fort, which assumed importance in Alfred the Great's time as a strategic vantage point from which to repel invasion by the Danes. These Danes occupied the Isle of Wight in AD 998 and ravaged many parts of Sussex for supplies. Outposts such as Warningcamp, just to the south of Burpham, were established to give warning of such attacks.

Burpham was the guardian of the Arun estuary well before Arundel Castle was built and the huge mound on which the original fort was placed, encloses some 20 acres just south of the village. In the spring its banks are covered with *daffodils* which make a wonderful foreground to the marshes and wildfowl reserve beyond. The name Burpham is derived from the words 'burh' or fort, and 'ham', village. It was mentioned in the Domesday Book as Bergeham, as was its ancient church, St Mary's. This church has its origins in the late Saxon period and has Norman arches, and a fine ribbed, vaulted chancel built in the Early English style about 1200 AD. As you walk through the south porch note the lovely chancel arch of alternate blocks of chalk and sandstone and at the appropriate season see the *swallows* nesting under the roof of the porch itself. This charming church with its protection of tall *conifers*, *holm oak* and *yew* leaves one with a sense of permanence and enduring peace. This feeling is heightened even more as you watch the evening shadows of a lowering sun lengthen and then slide subtly across the weathered stones of its western walls.

The road beside the church leads on to Peppering Farm where there is a bed of gravelly loam in which were discovered the bones and tusk of an elephant and some evidence of a primeval forest; now you may see domesticated Bison grazing in carefully fenced-off fields. Not far away was found the remains of prehistoric canoes. Above the road and just to the east of Burpham lies Perry Hill, where there are burial mounds which once contained Saxon bones. Three kilometres east of Burpham on the other side of Perry Hill once stood the twin village of Barpham which was completely erased by the plague in the fifteenth century. Every inhabitant died and all that remains today is a track with grass verges where the wattle and daub buildings stood, and a nettle-covered mound where the church once existed.

Just to the south-east of Burpham lies a wonderful expanse of woodland, Wepham Wood. Set in some of the most beautiful countryside in south-east England, this wood and its surrounding tracks combine extensive views of the open, rolling farmland, with the quiet intimacy of natural mixed woodland, all of which has something special to offer the naturalist in every season of the year.

Much of the woodland is made up of *English oak*, *Scots pine*, and *sycamore* but there are also areas of *beech* and *spruce*. In springtime the ground is a carpet of *bluebells* with *primroses, wood anemones* and *lesser celandines. Fallow deer* and *grey squirrels* can be seen, particularly in the north of the woods, and *badgers* are regular visitors although they are nocturnal by nature. *Adders*, too, are a common sight during the summer. The *white admiral* and many other species of butterfly can be seen in and around the wood and in the breeding season you should catch a glimpse of the *chiffchaff*, a migrant warbler from Africa. However, you are more likely to hear its characteristic song, two notes deliberately repeated in irregular order, 'chiff, chiff, chaff, chiff, chaff...' This bird, about 11 centimetres long, of slim and dainty appearance, is a characteristic bird of deciduous woodland, where it flits about the branches of tall trees, picking insects and their larvae off the leaves. Easily confused with the similar *willow warbler* also to be seen here, the chiffchaff has olive-brown upper parts and buffish white, tinged with yellow, lower parts. It has a yellowish stripe above the eye and has blackish legs, unlike the willow warbler which has light brown legs. The best way to distinguish between the two is their song pattern. The willow warbler has a more musical rippling sequence of notes, beginning quietly and become clearer and more deliberate, then descending to a distinct flourish 'sooeet-sooeetoo'. Also it is more characteristic of open bushy places and woodland clearings where it builds its domed nest of grass and moss, lined with feathers, on the ground and well hidden in tall grasses. The chiffchaff, however, builds its similar nest almost a metre off the ground in tall, rank grass supported by branches and small bushes; seldom is it built on the ground itself. Both these migrant warblers arrive in April and May and depart for tropical Africa south of the Sahara in August. Occasionally, and more frequently in recent years, the chiffchaff will spend the winter in Southern Britain and is particularly fond of suburban gardens.

It is time to return to Burpham and back to a little gem of a hotel, Burpham Country House Hotel. This is a good place to stay and is within easy reach and central to many beautiful parts of the South Downs. The hotel itself was once a hunting lodge for the Duke of Norfolk and a vicarage, and now provides a wonderful retreat from the stress of modern living to those who seek tranquillity and peace. Its restaurant provides, for me, one of the best menus of freshly prepared food I have tasted anywhere. Enjoy it all; the serenity of the surrounding countryside, the history, and above all a contentment and inner peace that will refresh you for your return to everyday life.

The setting sun across the water meadows of the Arun, highlights in stark relief, the tall imposing walls of Arundel Castle, where its surrounding woods and lake are the starting point of the next chapter.

Arundel Castle

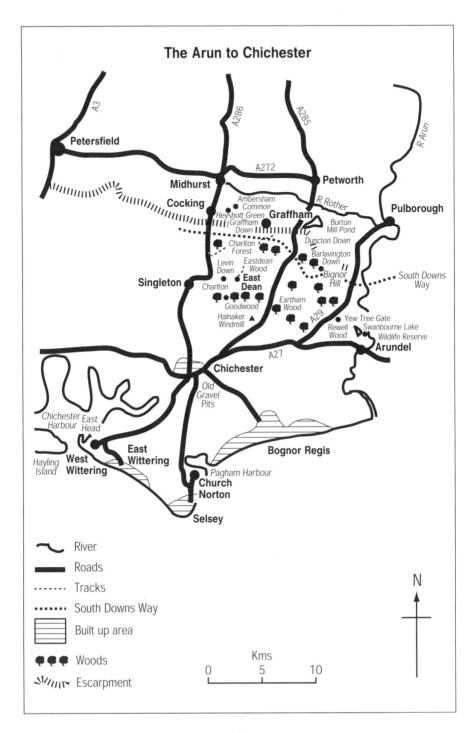

The Arun to Chichester

A3

A286

A285

R Arun

Petersfield

A272

Midhurst

Petworth

R Rother

Cocking

Pulborough

Ambersham
Common

Heyshott Green **Graffham**

Graffham
Down

Burton
Mill Pond

Charlton
Forest

Duncton Down

Barlavington
Down

Levin
Down

Eastdean
Wood

Singleton

Charlton

**East
Dean**

Bignor
Hill

South Downs
Way

Goodwood

Eartham
Wood

A29

Halnaker
Windmill

Yew Tree Gate

Swanbourne Lake

Rewell
Wood

Wildlife Reserve

Arundel

A27

Chichester

Old
Gravel
Pits

Chichester
Harbour

East
Head

**East
Wittering**

Hayling
Island

**West
Wittering**

Bognor Regis

Pagham Harbour

**Church
Norton**

Selsey

∿	River
▬	Roads
-------	Tracks
•••••••	South Downs Way
▦	Built up area
🌳🌳🌳	Woods
⬆🎏	Escarpment

N

Kms

0   5   10

148

7

THE ARUN TO CHICHESTER

Arundel Castle

Arundel is a short distance from Burpham by road, but by far the best way to approach is by the path from Warningcamp that takes you beside the river. The great walls of Arundel Castle are predominant and give a huge sense of indestructibility and permanence to the scene; their history is the history of England from the Conquest to the present. The town of Arundel too, has the same sense of quality and agelessness with decorous and dignified houses clustered all around its steep High Street. Arundel is also surrounded by all that is best in the English countryside; meadows, woods, quiet leafy lanes and hedgebound fields. Swanbourne lake and its associated wildlife reserve is our destination, less than a kilometre north of the castle itself.

Swanbourne Lake is pleasantly situated in a downland valley, enclosed with mixed woodland of *beech*, *sycamore* and *ash*. The lake itself can be traced back to before the Domesday Book when it was originally a mill pond used to supply water to the castle. Water for the lake collects from natural precipitation which seeps through the chalk to form springs at the head of the lake which are known as the Blue Springs. A constant check is kept on the water level to ensure that the lake never dries out, for the important wildfowl reserve downstream on the other side of the road is largely dependent on the condition of the lake.

The lake and its immediate surroundings are the home of 15 species of dragonfly, four species of bat, as well as numerous birds, butterflies and wild flowers. As you walk around the eastern shore-line look for a tall, somewhat lacklustre plant with yellow flowers. This is the *ploughman's-spikenard* which grows up to a metre tall along roadsides and tracks on dry, lime-rich soils throughout England and Wales. It flowers from July to October and has large, oval, lance-shaped leaves which are downy underneath and somewhat rough above. Spikenard is a very ancient name mentioned in the Bible and refers to a costly, perfumed ointment made from the roots of a Himalayan plant. This ointment, which had medicinal properties, was expensive to import, so poor people in Britain such as ploughmen, had to be content with a cheaper substitute from this indigenous species which they named plough-man's spikenard. When crushed the leaves and roots of the plant give out a sweet-smelling aromatic fragrance, and often they were dried and hung up in cottages as room-fresheners.

Harebells stand out amongst the grasses of the chalky banks beside the lake, their delicate blue colouring sometimes matching that of the early morning sky. The lake which is shallow in late summer, has the colours of brown and green in its waters. The north end of the lake narrows into an area of deep humus-rich bog, beloved by the *greater tussock sedge* which is prominent here with its substantial and dense tufts of razor-edged leaves. Nearby are areas of chalk grassland with good walks up through Arundel Park itself. A *green woodpecker* cries its loud and laughing call as it crosses the valley in its typical undulating flight pattern. A *kestrel* hovers in the distance just above the grass and then silently drops onto its prey; often you may see a *hobby* here displaying its acrobatic flight as it hunts for insects and small birds on the wing. This is a wonderful place to be which provides a respite of quiet and calm away from the hustle and bustle of the nearby town.

Tall *beech* trees surround the western side of the lake, giving dense shade and a deep cover of leaves on the steeply sloping hillside, and the bright green, shiny, strap-shaped leaves of *hart's tongue fern*, often 60 centimetres long, form large clumps all over the woodland banks. In one place, the roots of these beeches have been exposed over time by the elements, and seem to grapple the ground in all manner of shapes and sizes; almost as if the grotesque gremlins of *Grimm's Fairy Tales* had come to life. Nearby a small shrub with bright evergreen, oblong leaves catches your eye; this is *spurge laurel* which is common in South Down woodlands and grows beside paths and also deep in the woods themselves. Up to a metre and a half tall, this handsome plant is conspicuous at all times of the year, but it is at its best in the winter when its tubular yellow-green flowers and its shiny green appearance contrast so well with the pale grey of beech trunks and dark brown of decaying leaf-litter. The flowers are clustered about the upper leaves at the top of the stems and appear from December to April. During the day they have a musky scent which becomes stronger during the evening as they try to attract the year's first bees and moths, which may be about if the weather is unseasonably warm. Later in the year the flowers develop into black, oval fruits which like the rest of the plant are very poisonous. I was told an amusing story about the berries by a farmer whose house stood just below the heavily wooded northern escarpment. He was curious as to why his cats disappeared for several weeks at about the same time every year in the summer. One day as he was walking in the nearby woods, he noticed small heaps of spurge laurel berries with their skins neatly peeled away. He suspected that mice had been at work and so he kept watch one evening only to discover that his cats had the same idea, with an eye on the mice. Clearly when the berries had ripened sufficiently to attract mice, then the cats were happy away from home catching them. Interestingly, mice and the cats seem unaffected by the toxicity of the berries.

An exciting orchid to be found near Swanbourne Lake in late summer is the *green-flowered helleborine*, sometimes known as the *pendulous-flowered helleborine*. Compared to some of the other helleborine species, this one is small and somewhat delicate looking, with normally a single stem bearing up to 20 drooping green flowers, which are sometimes tinged with yellow or purple. The flowers do not fully open and often the upper ones do not open at all, remaining as tight buds. This peculiarity does not encourage insects to visit but is of little consequence because the flowers are always self-fertilised, fertilisation even occurring within those that remain unopened. The hairless stem grows from 10–40 centimetres tall, upon which the rounded and ribbed leaves are borne in two opposite rows, the largest leaves being above the middle of the stem. This is one of our rare orchids, so treat it with respect if you find it.

Before leaving the area of Arundel Park it is worth visiting the ponds and reed-beds of the nearby wildfowl reserve. Here, an impressive collection of wildfowl mingles with a variety of wild birds attracted by the food and lack of human interference. In winter, *moorhens, coots* and *water rail* feed among introduced and pinioned ducks of many varieties. There are several viewing hides which overlook the area and from November to March hundreds of wildfowl can be seen including *pochard* and *teal*. Waders such as *lapwing* and *redshank* can be seen at close quarters and in good years you may catch sight of rarities such as *bearded tits, whooper swans*, and possibly a *bittern*. A visit in late spring is often rewarding, for in addition to the captive wildfowl, you may also see the *little ringed plover* and hear the loud repetitive song 'settee, settee...' of *Cetti's warblers*.

The *little ringed plover*, with its distinctive yellow eye-rings, has an interesting history in Sussex. In 1949 two pairs bred in gravel pits near Chichester and up until 1955 some two to six pairs bred in the same locality. Subsequently, breeding was confirmed at Chichester Harbour and other places, but after 1960 the presence of nesting birds became irregular. Breeding success for this scarce migrant wader is not usually very high; possible reasons for failure being predation by *foxes* and *crows*, and heavy rain and cold weather at critical times. Of six pairs that bred in 1984, only ten young were raised and in 1986 five pairs raised eleven young. The little ringed plover like its close relation, the ringed plover, is a bird of undisturbed habitats and it prefers to breed on secluded gravel and sand pits and in the protected environment of nature reserves. At present, it continues to maintain its numbers and distribution, particularly in West Sussex, where there are now perhaps about fifteen breeding pairs scattered about in several localities. The best time to see the bird is during the spring or autumn migrations when it can be seen feeding with other waders on mudflats, seashores and wet places inland.

Rewell Wood, just to the west of Arundel, is our next destination and it is best approached from one of the several lay-bys on the A29. Numerous tracks climb up the grassy surrounds of the woodland edge on this western side and lead you up through the wood itself. The grassy slopes are well worth investigating as they are covered with typical chalk-loving plants, including *wild basil, common St John's-wort, felwort* and *harebell*. Quite large areas are covered in *marjoram*, which here is seen also with pure white flowers as well as the more normal purple. Marjoram grows up to 60 centimetres high and can be easily distinguished by its egg-shaped leaves and closely packed heads of tiny, aromatic scented flowers. It blooms from June to September and attracts many species of insects and butterflies. The generic name *origanum* is derived from the Greek *oros* a mountain, and *ganos*, joy, describing the colourful appearance of the plants growing on the hillsides. The ancient Greeks believed that marjoram on a grave signified the well-being of the departed soul. Known also as oregano, marjoram is a characteristic herb of Mediterranean cooking. In Kent, it was once gathered and hung up to dry in cottages and then used to make tea. It was also used medicinally in the sixteenth century, when the dried flowers were chopped up and added to a jar of sugar left in the sun for 24 hours, this sugar being taken over a period of a few days to cure diseases of the eyes and kidneys.

A *gatekeeper* butterfly, with distinctive false eyes on its orange and brown wings, joins others like the *marbled white, painted lady* and *red admiral* as they gather around the strongly-scented marjoram. The gatekeeper is a butterfly of high summer which likes to bask with wings spread in warm corners along hedgerows and woodland paths. It is one of the most abundant butterflies where there are grassy places with nearby cover of shrubs or trees. It is often confused with that much more common butterfly, the *meadow brown*, but the difference between them is quite obvious, the gatekeeper being smaller, and having two white pupils in the black eyespot on the upper and lower sides of the forewings. Gate-keepers are territorial, and like the *speckled wood* species, they patrol their areas of hedgerow and track to ward off other insects. The female, which is larger than the male and lacks the dark band of scent scales across the fore-wing, lays her eggs singly in August amongst a variety of common grasses. After three weeks, the cater-pillar, coloured brown with white stripes, emerges and feeds on the grasses. It hibernates over winter, then feeds again in the spring, before spending a further three weeks as a chrysalis. The first adults emerge in late June to be followed by others in July and August; they live for only about two to three weeks.

The day is hot and the sweet scent of dog-roses is in the air as you climb up the hillside towards the wood. The *dog rose*, because it is so ubiquitous in the British countryside, is often taken for granted. However, it is only one of a vast multitude of species, subspecies and varieties of wild rose here in Britain, some of which have been used by horticulturists in the evolution of various groups of cultivated roses. The dog-rose forms a large bush up to 3 metres high, with long arching branches, covered with broad, sharp thorns. The leaves are made up of five leaflets and the five-petalled, usually pink flowers, blossom in June and July. Wild roses form colourful scarlet or orange fruits, oval in shape, called hips which adorn our hedgerows in autumn. These hips are very rich in vitamin C, and during and just after the Second World War, they were collected by the public and sold to chemist's shops. Hundreds of tons of hips were collected annually and were made into rosehip syrup using a recipe given by the Ministry of Food in their booklet *Hedgerow Harvest*. This syrup was sold at a controlled price through chemist's shops, and mothers of young children could obtain it at reduced prices through welfare shops. Rosehip syrup is still made commercially today and there are many modern recipes that make use of it in the kitchen. Interestingly, in Roman mythology, Cupid is said to have given Harpocrates, the god of silence, a rose to bribe him not to betray the amours of Venus. For this reason the rose became the emblem of silence and was carved on ceilings of banqueting rooms to remind guests that anything spoken with wine at the table was not to be repeated outside. From this was derived the expression sub rosa, literally 'under the rose', meaning in strict confidence.

As you climb the hill into Rewell Wood, the noise of the traffic on the busy road below becomes less pronounced as the enveloping branches of the trees close around you. The *spindle tree* described fully in Chapter 3 can be found here in plenty, as can the *buckthorn* shrub. Buckthorn grows up to 3 metres high and is found chiefly in hedges and woods on chalky soils in southern England. Its spreading branches grow opposite each other, the smaller ones terminating in sharp thorns. The bright-green leaves are about 3–6 centimetres long, egg-shaped and have prominent veins. The small yellowish-green flowers appear in May and June and are followed by shiny blue-black globular berries in September. These berries have been used in the past by herbalists as a purgative in the form of syrup which considerably modified their drastic action if used in their natural state. The *purging buckthorn*, as it is also known, has other uses; its bark yields a yellow dye and the juice of its berries provides a green pigment used by artists.

The trees are taller at the top of the hill and consist mainly of *sweet chestnut*, *beech*, *larch* and *ash*. In places when there is a reasonable depth of heavy clay soil above the underlying chalk, good specimens of the *common oak* are evident. As you walk underneath these lovely tall trees in late spring, you will almost certainly hear the sharp, rather loud call 'chwit, chwit, chwit' of the *nuthatch*. There are several pairs nesting in Rewell Wood and in the winter months, listen out for its tapping as it uses its sharp, strong beak to drill out the kernels from nuts which it has skilfully wedged into the crevices of bark. Only some 15 centimetres long, the nuthatch is distinguished by its short tail, blue-grey upper plumage, bright buff underparts, with a bold black eyestripe, and white cheeks and throat. It climbs up and down trees in short jerks and is the only British bird that regularly moves downward, head first. The nest of the nuthatch is made of bark flakes or dead leaves, in a hole in a tree, often one that was once used by a wood-pecker. Remarkably, the nuthatch plasters up the entrance to the nest with mud, reducing it to the bare size necessary, thus ensuring that unwanted intruders are kept out, and in April and May, six to ten white, red-spotted eggs are laid. The young are fed on insects which the parents have searched for under tree-bark. Nuthatches spend the winter foraging the woodland, often joining flocks of tits looking for nuts, acorns and beech mast. They are particularly susceptible to the cold and many perish in a hard winter. Often in harsh weather, they will seek food from garden bird tables and if a suitable nest box is available a pair may well choose to use it in the spring.

Follow the paths along the top of the wood in a north-easterly direction which soon opens up at a place well-signposted as 'Yew Tree Gate'. Here, huge *beech* trees send branches high into the sky competing with other tall trees of *oak* and *ash*. An ancient boundary-bank trails across the ground beside a wide, deeply rutted trackway, for this was once just part of a great commercial forest used for the coppicing of *sweet chestnut* and *hazel*; the tracks are a reminder of those operations.

The ditches alongside the tracks radiating out from Yew Tree Gate are damp, and the ruts often contain water. Here you will find the *corn mint* and the *yellow pimpernel* growing. Corn mint grows from 10 to 30 centimetres high and has stalked, elliptical, blunt-toothed leaves. The lilac-coloured flowers bloom from April to October and are clustered around the leaves below the top of the stem. Unlike most other species of the mint family, the corn mint does not have a minty smell when it is crushed between the fingers, being more acrid.

The *yellow pimpernel* like the corn mint is found in damp places, in woods and copses, sometimes in large patches. It is a neat, hairless plant with oval-shaped leaves that grow opposite each other on the stem. The whole plant trails along the ground and has solitary bright yellow five-petalled flowers which are held on slender long curved stalks, and appear from May to September.

In the southern part of Rewell Wood there is a good area of *sweet chestnut* coppice which is frequented by *nightjars*, especially after the wood has been cut. The nightjar with its wonderful camouflage of mottled brown plumage against a background of dead leaves and wood, is seldom seen during the day, when it sits motionless along a branch or on the ground. Just before nightfall, however, the male begins a nocturnal 'churring' song which lasts for as long as 15 minutes or so. The female then joins up with the male and they fly off to hunt for large moths that form the major part of their diet. It is in the evening, then, that you are more likely to spot them flying over the wood, silhouetted against a pink sky. The nightjar is migratory, arriving from Africa in late April and departing in September. It is a bird of sandy heathlands and chalky soils both of which provide free-draining ground for the nest site. The two white eggs, blotched with brown markings, looking like beach pebbles, are laid on bare ground in May. The nest site can be under bushes, in nettles, or even in the open. Both parents look after the young, the male often finishing the task when the female commences a second brood.

Another bird you may come across in the coppiced wood areas of Rewell Wood is the *woodcock*. This bird is about 33 centimetres long and has a prominent long stout, thick bill. The colouring of its upper parts with black, brown, red-brown and buff, together with the creamy and finely barred underparts, give it perfect camouflage that blends in so well with the play of light and shade on the dead, brown leaves of the woodland floor. Observation of this bird is made difficult, not only by its superb camouflage, but also by its shy and retiring nature. You are most likely to see it in the spring when the male woodcock defends its territory with a 'roding' flight. In this, it flies above the trees round and round its domain with its bill pointed downwards, and uttering a soft croaking 'orrrt-orrrt', repeated three or four times, followed by a thin sneezing 'tsiwick'. It makes its nest of dried grass and dead leaves in a natural dry hollow in the ground. The four handsome eggs of pale buffish-red, blotched with ash-grey and brown, are laid in March and April, with repeat clutches a month or so later. Both sexes take turns in incubating the eggs and when the young leave the nest just a few hours after hatching, they are looked after by both parents until they are able to fend for themselves.

The woodcock in Britain is both a resident, a winter visitor from the north, and a passage migrant. It is a solitary feeder seeking worms at night by probing worm-rich grasslands in woods and open downlands with its bill. You may see it flying to its favourite feeding places on the Downs at dusk which it does regularly, or you may disturb it during the day when it lies up in the patches of scrub and woodland. I hope you will enjoy the satisfaction of seeing and hearing the woodcock not only in this wood, but in many of those further to the west.

A butterfly of the newly coppiced areas and sunny rides in Rewell Wood is the *pearl-bordered fritillary*. This species has orange-brown wings with dark spots, and two silver patches on the underside of its hind-wings. Sadly, it has rapidly declined throughout Britain in the twentieth century. This decline is almost certainly due to the ploughing up of the meadows and glades, its natural habitat, and more recently by the widespread use of herbicides that has destroyed many nectar-rich wildflowers like the *bugle* and *yellow archangel*, and its caterpillar food plant, the *violet*. The pearl-bordered fritillary is now restricted largely to woods and coppices where various species of violet still grow in clearings and along paths. The butterflies spend much of the day patrolling their territories and in the evening they will bask in the setting sun before resting for the night under the heads of grasses. The best time of the year to see this attractive butterfly is from early May to the end of June.

Rewell Wood is also the home to many types of fungi and when the first autumn rains begin and the leaf litter becomes damp, then fungal mycelium begin their work and there is an aroma of decay. This damp smell is not caused by the rain but is the result of millions upon millions of tiny spores of various fungi. Two species you will find at Yew Tree Gate are the *oyster mushroom* and the *beefsteak fungus*.

The *oyster mushroom* has a smooth shell-shaped cap 6–14 centimetres across. It can be found at any time of the year but is more common from the end of summer through to winter, growing in large clusters on stumps of old deciduous trees, especially *beech*. The colour of the cap varies from flesh-brown to a beautiful deep blue-grey, and its gills are white at first and later having a yellowish tinge. The whole fungus has a pleasant smell and taste and is well worth cooking in the same way as you would the *field mushroom*.

The *beefsteak fungus*, so called because when cut up, its flesh looks like steak and it exudes a reddish juice. It is one of the bracket fungi which grow out of the wood of live *chestnut* or *oak* trees, as well as those that are dead. They have small crowded tubes below their cap instead of gills and often form extensive patches with the brackets arranged in tiers along the trunk. The beefsteak fungus is 10–25 centimetres across and coloured pinkish to orange-red. It has a pleasant smell and despite its sour taste it is delicious when fried in butter with finely-chopped shallots, garlic and other herbs. This fungus causes brown rot in wood and when oak timber is infected it attains a much darker and richer colour than normal and this 'brown oak' is much in demand and valued for furniture, after suitable treatment of course.

Just about anywhere in the countryside or in the home you will come across *spiders*. There are 600 species of spider in Britain, all of them harmless to man, if you ignore the occasional visitor from foreign lands like the *tarantula* in a cargo of bananas. Spiders have eight legs, a body divided into two, a head and thorax separated by a waist from the large abdomen. Most have eight eyes arranged in two rows of four and all prey on other creatures, often by spinning webs of silk strands which, thickness for thickness, have a breaking strain greater than iron. The thread is extremely thin and is so light that were a single strand to be strung around the world it would weigh less than 200 grams. Not all spiders produce webs; some catch their prey by leaping upon them, while spitting spiders discharge streams of gummy thread that ensnare their quarry. However, all spiders finally kill their victim by stabbing them with fangs which inject poison. Spiders belong to the class Arachnida which is one of the oldest on earth with a history stretching back to as long ago as 350 million years in the Devonian period, and includes the *scorpion*, *tick*, and *mite* species. The class is named after the mythological maiden Arachne, who challenged the goddess Athena to a weaving contest and was transformed into a spider. Hence the belief that it is unlucky to kill a spider.

One of the spiders to be found on earth banks in Rewell Wood is the *ant spider*, so called because its movements, size and general appearance are antlike, and because it is often found running close to ants. It is only about 4 millimetres long, about the same size and with the same shape as a red worker ant. The ant spider, when seen with a hand lens, is a beautiful little spider, having scale-like hairs which scintillate in the sun with all colours of the rainbow. This spider is a night hunter, preying on other small insects, and when not in movement and therefore, not in danger from enemies, it encloses itself in a silk cell under a stone or piece of wood. The female's eggs are contained in thick white egg sacs which are encased in the spacious silken cell. Young spiderlings hatch out in the spring and after a few days are able to scatter to lead their own lives. Spiders have a lifespan of about a year.

Closely related to the spider is the *harvestman* of which there are 26 species in Britain. They all have eight legs like spiders, but can be distinguished from them by having the head and thorax attached to the abdomen without a dividing waist, giving a one-piece body. They feed on insects and decaying plant material and they do not trap prey in webs of silk. Harvestmen avoid activity during the day and like to bask in the sun; at night they scuttle over vegetation in search of food. They live from four to ten months. During the early summer in Rewell Wood, you are quite likely to see a species of harvestman called *megabundus diadema* on a tree above ground level. The body of this species about 5 millimetres long, is often brightly coloured and its eyes are set on top of a spiny 'turret'.

As you descend through the wood on your way to Bignor Hill look out for the very rare hybrid between the *fly orchid* and the *bee orchid* discussed in the last chapter. This remarkable orchid has some of the characteristics of the fly orchid, especially its colouring and also displays the broad concave labellum and elliptical sepals of the bee orchid. June is the best month to see this plant which has only been discovered in the last few years, but similar forms could be found in just about any place where both individual species grow in proximity.

As you approach the broad smooth slope of Bury Hill, having travelled north from Rewell Wood, you will have passed across, or near to, the huge linear earthwork of War Dyke. Over 18 metres wide in places, and double banked as it passes through the woods towards Fairmile Bottom, this great work is prehistoric and may have formed the ancient trackway from Chichester to Shoreham. There are other earthworks in Rewell Wood with tumuli and also the remains of ancient settlements, so this whole area was once of great importance whose origins are lost in the mists of time. Was there once an ancient Celtic city on Rewell Hill?

From Bury Hill, rising as it does out of coloured woods of *oak*, *silver birch*, *beech* and *bracken*, the South Downs Way takes us next on to the top of Bignor Hill with a neolithic campsite at Barkhale Down just to the south. The great Roman road of Stane Street lies just to the west, with its associated Roman villa a short distance away, to the east of the village of Bignor. This impressive villa is worth visiting as it is one of the largest ever found in England, covering several acres, and includes Roman baths with their heating system, and a beautifully decorated floor with coloured pictures, Roman craftsmen using chalk for the white, bricks for red, sandstone for buff, and limestone for blue and green. The subjects, from Greek mythology, depicted by these mosaics, are typically Roman in style; Ganymede being carried off by the eagle and the head of Medusa complete with its snakes.

The sides of Stane Street as it climbs up the side of Bignor Hill are well wooded and there are good colonies of *common spotted* and *bird's nest orchids*. As you reach the highest point of this old Roman road, the top of Bignor Hill is but a short distance away. Nearby is the neolithic causewayed camp, a number of Iron Age defensive dykes, and several Bronze Age burial mounds; these all give a deep feeling of history and permanency, and yet at the same time you have an awareness that man himself has passed by just fleetingly.

The views from Bignor Hill are what you have come to expect from the top of the escarpment of the South Downs. To the north, the Weald as ever with lowlands of dim blue, beyond which green hills stretch across the horizon giving a sense of freedom, space and a certain remoteness. To the east over cultivated fields are the distant downs with the bare folds and bluffs that you have come to know so well in your travels.

To the west, the landscape is transformed into an immense spread of rolling, wooded country, and only in small areas is there a reminder of open chalky grasslands. Many of the woods are of *beech*, planted about two hundred years ago, particularly on the escarpment where they now form huge hangers. The western woods were well known to that poet and historian Hilaire Belloc who lived nearby at Slindon. In his poem 'The South Country' he catches the mood of these places.

> The great hills of the South Country
> They stand along the sea.
> And it's there, walking in the high woods,
> That I would wish to be,
> And the men that were boys when I was boy
> Walking along with me.

These extensive woods are the home for two of our birds of prey, the *sparrowhawk* and the *goshawk*.

The *sparrowhawk*, because of its shy and secretive habit, is not as well known as the more familiar *kestrel*, which is so often seen hovering along our roads and motorways. Apart from the *buzzard*, the kestrel is the only British bird of prey that does hover. The sparrowhawk does not hover and can be distinguished from other small birds of prey by the combination of short, rounded wings and long tail. It has long yellow legs, and fierce yellow eyes which strike terror into small birds which are its main source of food. It catches these after a sharp fast flight from some unexpected quarter such as when it flies low along a woodland ride in the morning or late evening and then makes a sudden dart into the midst of a flock of feeding finches. I remember one cold, late autumn evening when looking down into my back garden I saw all the small birds at the feeding table suddenly take fright and fly swiftly into the surrounding hedges. Almost simultaneously a sparrowhawk flew at great speed, low across the garden; it was amazing to see the speed of reaction of the hawk's intended prey, as instinctively they knew when danger threatened.

The sparrowhawk makes its nest high up in the top of the trees close to the main trunk in the form of a large platform of sticks and twigs. The five or six white eggs are beautifully blotched with pale and dark rich brown, mostly towards the larger end, and are laid from April to June. They hatch out after about 35 days and the adult male then does all the hunting for his mate and young, bringing small birds, still feathered, to the nest. The female tears them into pieces and feeds the bits to the young until such time that they can do it themselves, usually when they are about three weeks old.

At one time the sparrowhawk was one of our most abundant birds of prey, but as woodlands were felled in the nineteenth century, so their population decreased. However, this clever and cunning bird survived in remoter and undisturbed woods and forests, and even persecution from gamekeepers did not seriously deplete their numbers. It was not until the 1950s and 1960s, following the introduction and widespread use of pesticides, that the sparrowhawk numbers started to seriously decline. Its breeding capability was particularly affected by DDT which, unlike other poisons, is extremely stable and its toxicity stores up and remains in the body of its victims and affects a large proportion of the food chain. Since 1964, successful efforts have been made to control the use of toxic chemicals by farmers, especially DDT, and the sparrowhawk has made a remarkable recovery.

The *goshawk* is a much rarer bird than the sparrowhawk and although it has the same colouring of grey-brown with a barred breast, it is very much larger. The goshawk is a bird of large woods and forests, each pair requiring some 8,000–12,000 acres, and in Roman up to mediaeval times, when much of Britain was covered with huge continuous areas of deciduous and coniferous forest, this bird together with the sparrowhawk must have been relatively numerous. However, by the early nineteenth century more than ninety per cent of Britain's forest had been cleared and what remained was in scattered areas, individually too small to support pairs of goshawks; therefore, by 1800, this species became extremely scarce and very localised. The era of game preserving then came, the last few remaining goshawks were pursued and killed and by 1900 the goshawk was virtually extinct in Britain. Although goshawks have also declined elsewhere in Europe, it is possible that since the Second World War, some natural colonisation has taken place from that source. Certainly in recent years there have been more and more sightings of the bird, particularly in West Sussex. Interestingly, a pair was observed in 1976, where the male was displaying and carrying food, a sure sign that they were breeding. The display flight usually takes place in the early morning and is similar to that of the sparrowhawk, with soaring, slow wing-flapping, and some shallow undulating flights. The nest is a large structure of sticks high up in a woodland tree and the one to five eggs are laid in May, hatching out in June, the young being able to fly about the middle of July. The goshawk hunts for birds, especially young *wood pigeons* by a swift, dashing flight, fast and low among trees, with a few rapid wing beats and long glides, having an amazing agility to avoid trees, doubling past each one, in its pursuit through the forest.

In the Middle Ages when falconry was the great sport of country gentlemen, the goshawk was one of the hawks that was used. It was trained to catch rabbits, partridges and even hares. It was said of falconry at that time, that the useless *kestrel* was for the poor man, the *sparrowhawk* was suitable for the priest, the *goshawk* for the squire and the *peregrine* for the nobleman. Falconry is today quite popular and it is possible that escapes from this source have established themselves in the wild. What a great benefit to our heritage it would be if this wonderful hawk was once again a regular breeding species and it was able to increase in numbers in our southern woodlands. Perhaps you will be lucky enough to hear its short plaintive mewing 'kew-kew-kew' or its chattering 'gig-gig-gig' in your walks through remoter parts of the West Sussex forests.

The woods below Barlavington Down and Duncton Down contain many species of interesting plants. Two which are particularly abundant here are the *bird's-nest orchid* and the *nettle-leaved bellflower*.

The *bird's-nest orchid* is 20–50 centimetres tall and is all the same colour, a warm yellowish-brown. The stems are sheathed in long brown scales, being devoid of leaves. The single flowering spike carries 50–100 large flowers which appear in May and June. These flowers have a pleasant scent of honey emanating from the nectar which forms in a shallow cup at the base of the labellum and this

162

attracts many types of small flies which then act as pollinators. This species flowers in the darkest part of the woodland and here at Duncton it is most common at the bottom of the old chalk excavations at the sharp bend on the eastern side of the A285, opposite the viewpoint car park. The root system is a tangled mass of fleshy fibres giving some semblance of an untidy bird's nest and is infected with mycorrhizal fungus which breaks down dead beech leaves to give the orchid nutrition. It is therefore wholly saprophytic, not requiring sunlight to produce its food by photosynthesis. You should have little difficulty in finding this orchid for it grows in most woodlands on the chalk of the western portion of the South Downs.

On the steep banks of the woodland paths in July to September you will not fail to notice the 60–90 centimetre tall, elegant-looking member of the bellflower family, the *nettle-leaved bellflower*, also known as '*bats-in-the-belfry*'. This plant has a stout, unbranched, bristly, sharply-angled stem which supports large violet-blue bell-shaped flowers in clusters, those at the top being the first to open. Its irregularly toothed leaves are heart-shaped at the bottom, becoming more lance-shaped nearer the top, and before flowering the whole plant has a nettle-like appearance.

Well worth a visit is the viewpoint situated beside the main road as it climbs up the flank of Duncton Down. The Weald and the North Downs with Blackdown, Bexley Hill and Leith Hill stand proud across a far countryside of fields and woods; small Sussex villages, looking peaceful and unperturbed with their church spires or bold Norman towers pointing upwards to the sky. Immediately below in the beech hanger lies Duncton Chalk Pit Nature Reserve, and nearby is another nature reserve at Burton Mill Pond. Both reserves are managed by the Sussex Wildlife Trust, the former being leased to the Trust and the latter being owned by it.

At Duncton chalk pit is an old chalk working with the ruins of a lime kiln together with an associated access track. Until the middle of the nineteenth century there were a number of limekilns in operation in Sussex where chalk was burnt to produce agricultural lime for spreading on fields. The skilled craft of the lime-burner was handed down in families from one generation to another, and specially built brick furnaces were constructed in which gorse, broom and wood were burnt as fuel. This kiln at Duncton was once an unusually large enterprise owned by Lord Egremont of Petworth, until competition by limeworks built on the waterways of the nearby river Rother caused it to be closed. The trackways around the chalk pit form a contrasting area of sunlight on air, and in the open parts and amongst its scrubby margins can be found more than 150 plant species, including numerous orchids. One such species, the *wild liquorice*, is particularly uncommon.

The *wild liquorice* is also known as *milk-vetch* because of the initial sweet taste of its roots and leaves, which soon goes, leaving an unpleasant flavour in the mouth. It is a conspicuous sprawling plant with stems up to a metre long, and leaves without tendrils, each with four to six pairs of oval leaflets, the broadest of which are in the middle. The greenish-yellow, pea-like flowers appear in July and August and are arranged in a short spike up to 5 centimetres long. This plant should not be confused with the true liquorice plant, which is a native of south-east Europe and western Asia and was once cultivated around Pontefract in Yorkshire for the liquorice industry.

The reserve at Burton Mill Pond consists of four wooded areas, all derived from the *oak* and *birch* woodlands typical of the sandstone parts of the Weald. The acidic nature of the earth here creates many heathy areas where trees were cleared in the past. Three species of the heath family can be found flowering in the summer on these acid soils, namely, *ling*, *bell heather* and *cross-leaved heath*. The most common is the heather, or *ling* as it is usually called, with its opposite leaves closely packed on the twiggy stems along which grow the numerous small lilac-coloured flowers. The Latin name for ling, *calluna*, gets its name from the Greek word *kalluno*, meaning to beautify, and those who appreciate the beauty of our moorlands covered with this plant will agree with this apt name; in reality, however, the name was given because this heather was made into brooms which did the beautifying by sweeping.

Bell heather grows on the drier parts appearing side by side with ling and is easily recognised by its reddish-purple flowers, shaped like small bells, which are clustered around the end of the woody stems. The *cross-leaved heath*, on the other hand, favours the wetter and more boggy parts, and unlike the bell heather, is covered with soft hairs and has relatively large rose-pink flowers in terminal clusters which 'nod' in the full flowering stage. As its name implies its leaves are arranged around the stem in fours, like crosses.

The boggy areas, like the aptly named Black Hole and Welch's Common, can be crossed by a boardwalk from which can be seen *bogbean* and *marsh cinquefoil*. The bogbean puts out spikes of pink and white flowers in May and June, but for most of the year, it is recognisable only by its large clover-like leaves up to 10 centimetres across which rise above the water surface or on top of the boggy areas. The leaves were once used instead of hops to give a bitter flavour to beer, especially in northern England. Together with the leaves, the rhizomes contain bitter compounds and were used as a tonic, as a sedative, and also to aid digestion. In ancient times the plant was held to be of great value against scurvy.

Marsh cinquefoil is quite distinctive, as many parts of it are purplish in colour, including the large five-petalled flowers which appear in June and July. This cinquefoil has a long woody, creeping rhizome and more or less erect stems. The lower leaves have five to seven oblong leaflets and the upper ones have only three. This is an uncommon plant in southern England and is particularly vulnerable to change or destruction of habitat. Fortunately, this reserve is carefully managed in order to keep the heathy glades and bogs from being completely overrun with *alder* and *willow* which would soon lead to the disappearance of the marsh cinquefoil and other species.

In the drier parts of the woodland on the reserve the *birch tree* is too dominant so some areas of these have been replaced with a blend of broad-leaved trees such as the *oak*, mixed with *crab apple*, *wild cherry* and *spindle*. This has created a more natural woodland with open areas beloved by many butterflies.

The huge Burton Mill Pond dominates the central area of the reserve and this of course adds another habitat for wildlife. This large body of water was created in the sixteenth century for use as a hammer pond in the iron industry, overlooked on the north side by a converted watermill dating from the eighteenth century. The best way of viewing this large stretch of water is by standing beside the minor road that passes over the causeway to the north. Here the placid waters have a wonderful backdrop of steeply sloping downland clothed with trees. Reeds soften the edges of the lake and give cover and nesting sites for *coots, moorhens* and *great crested grebes*. This last-named bird is the largest of the five British grebes and can be easily identified in summer by its black cap, and prominent chestnut and black head plumes, which are erected when in display. Grebes pair for life, but they still have complex courtship rites in the early spring. The couple swim towards each other, then their beaks touch and they shake out their head plumes displaying all their beauty; at the same time they waggle their heads to and fro with ear tufts erected. At times, both birds dive and bring up weed which they offer to each other as they pose for a few seconds, their breasts almost touching and each swaying from side to side. They then part, only to repeat the performance once again. These grebes are masters of the water and are expert divers seeking fish, frogs and tadpoles under the surface. Watch them swim slowly across the sparkling water and then effortlessly slip beneath the surface disappearing for fifteen seconds or so, only to bob up again several metres away. Occasionally you may see one take to the wing with a lengthy pattering of feet on the top of the water before becoming airborne.

The nest of the great crested grebe is made of many kinds of dead water plants and usually floats on the surface amongst reeds and overhanging branches, but occasionally its foundation rests on the bottom. The three to five, usually four, white eggs are laid from March to June and are covered over whenever the adult leaves the nest.

This grebe is not silent by any means, and often in the quiet of the evening you will hear rattling and trumpeting cries, sometimes a barking 'Kar-arr' and sometimes a shrill 'er-wick', accompanied with other similar moaning and whining noises. The great crested grebe is a resident species in the country and you will see it here all the year round. It has been increasing over recent years, and judging by the numbers of little black and white striped chicks seen in late spring or early summer, this increase is being sustained.

Unlike the conspicuous grebe, the *water rail*, which also breeds around the Mill Pond, is a highly secretive and shy bird and you are more likely to hear its pig-like grunting and squeaking, at night. It is similar in appearance to a *moorhen*, but with a longer red bill, a greyer breast and with barred black and white flanks. You may get a clear view of this attractive and graceful bird when it occasionally emerges from cover and swims amongst the waterside vegetation or seeks insects and earthworms on open ground.

Before leaving the wonderful nature reserve with its distinctive habitats, take the track that leads north-westwards from Crouch Farm. This quiet way leads down through a sheltered and shallow defile with low sandstone cliffs on either side, to an old dam that divides Burton Mill Pond from another artificial lake higher up to the south-west. This peaceful and tranquil place is a haven for wildlife. A small stream trickles over the old dam dividing the two large stretches of water; nearby large areas of thick mud lie exposed with *greater tussock sedge* growing upon it. Bees hum in the still warm air of an early summer day, and *damselflies* with weak, fluttering flight, seek small insects on which to feed amongst the lush vegetation. Two butterflies are common here, the *green-veined white* and the *speckled wood*.

The *green-veined white butterfly* is easily distinguished from other white butterflies by the prominent greenish-yellow veins on its wings which are particularly noticeable on their undersides. Even on dull days the butterfly is active, and is frequently seen in damp places near hedgerows and beside woodland edges. The green caterpillars, distinctive with yellow marks around their respiratory openings, feed on wild plants of the cabbage family such as *hedge mustard, garlic mustard, charlock* and *wild mignonette.*

The *speckled wood* likes to settle where dappled sunlight percolates through the leaf canopy, for in such places its brown and whitish-buff speckled wings act as a fine camouflage. The male, which has smaller spots than the female, settles in patches of sunlight where a female is more likely to see it. This species shows a high degree of territorial behaviour, and the male will defend its patch by attacking any intruder. Sit here in this secluded place under the shade of *holly* trees and beside tall purple *foxgloves* and watch a male speckled wood fly in a series of up and down movements to harry anything that dares encroach onto its territory. The two butterflies tussle together, spiralling upwards in their mutual anger but it is the defender that eventually succeeds in seeing off the other. As with most butterflies, scent produced from the base of the wing scales, plays an important part in the speckled wood's lifespan of only 20 days. The male almost certainly marks its territory by scent which is also a possible attraction to the female. The eggs are laid twice a year in various grasses any time between spring and autumn; the small green caterpillars emerge in 10 days and then feed at night on the grasses for about 30 days. They then turn into chrysalids, before new butterflies emerge a month later. There are usually two generations a year in Britain and caterpillars that hatch in the autumn live throughout the winter, actively feeding on warm days.

In places around the margins of Burton Mill Pond and growing on the virtually floating masses of fen-like peat, is a very rare species of the carrot family, *cowbane.* It can be distinguished from most other aquatic members of the same family by its hairless, hollow stems, by its large size, up to 150 centimetres tall, and by its leaves, which are much divided into narrow saw-toothed segments. In addition these are no bracts or small leaves below the lower level of the umbel stalks. The tiny white flowers, only 3 millimetres across, bloom in July and August. You should treat this plant with caution as it is extremely poisonous, due to a resinous substance produced in the yellow juice of the roots and stems.

Before returning to the Downs, it is worth visiting Ambersham and Heyshott Commons 7 kilometres or so to the west. These two small nature reserves are managed by the Sussex Wildlife Trust and consist of heathlands on the lower greensand rocks, just south-east of Midhurst. Amongst the vegetation of *ling* and *cross-leaved heath*, and making particular use of open sandy areas, is a wide variety of heathland invertebrates, including many beetles, bees, wasps, spiders and flies. Many of these are found only in the warm, sheltered conditions of heathlands in the south of England.

Wolf spiders of various kinds are particularly fond of these dry heaths and they are a species that do not make a web to catch their prey. Instead, they are often seen scurrying across bare ground chasing their victims which when caught, are quickly killed or at least paralysed, by a venom injected through the wolf spider's powerful fangs. The prey is then sucked dry of all its juices and left as a husk. The courtship display of wolf spiders is as varied as it is elegant and in some cases the male will offer the female a gift of a dead fly wrapped in silk, and at the same time as she eats it, he mates with her. Without such a gift, she is quite likely to eat him! Most wolf spiders lead a nomadic existence and because of this they have to carry their egg-sacs around with them under their bodies, until they are ready to hatch. Even then, some species carry the young spiderlings on their backs for a week or more. Wolf spiders are generally quite large and can be recognised when they are basking in the sun by the way the two pairs of front legs are held quite close together. Enjoy looking for these spiders on the heath and although one of the largest found here is known as *tarantula barpipes*, it will not harm you.

Adders and *common lizards* are often seen basking on warm days and many birds like the *nightjar*, *stonechat* and *tree pipit* are attracted here by the variety of insects. My last visit to Heyshott Common was in late spring 1999, and I was lucky enough to see and hear an uncommon bird, the *woodlark*.

The *woodlark* is similar in appearance to the *skylark*, but is a few centimetres smaller and has a shorter tail, which has no white sides to it. This is a bird of bracken and bramble-covered heathland where there are some surrounding trees. It often perches on trees and bushes where it puts out a wonderful and melodious 'toolooeet' interspersed with a sequence of trilling notes 'lu-lu-lu-lu'. Its song is sweeter, but less sustained and less powerful than that of the skylark, yet many believe it to be more musical than that of its well-known relative. The male also sings its song while in flight, often rising in the air on rapidly fluttering wings, circling widely and then spiralling upwards before at last descending, either gradually, or with an abrupt plunge similar to that of the skylark.

At one time in the 1950s and 1960s the woodlark was quite numerous on Sussex heathlands but over the intervening years it has declined for some unknown reason, and in recent years there have been only a few breeding pairs, mainly in West Sussex. Encouragingly there is now some evidence to show that the species has begun a slow recovery in numbers.

As you leave Heyshott Common look back and see the tall *Scots pines*, like sentinels watching over the wide heath, and watch the pale green leaves on the whip-like branches of the *silver birch* shaking in the breeze. See, too, the pale greenish-grey tussocks of *purple moor grass* as they catch the light of the sun and sky and sway to and fro amongst the darker heathers. Out of place in this natural setting are ugly black power-lines which pass overhead, a reminder of how the essentials of man must intrude at times on the natural world.

Heyshott Green with its small pond is just a short distance away and in June the Green here is covered with *heath spotted-orchids* and the rose-red flowers of *ragged-robin*. This last-named is a familiar plant of damp places, where its unmistakable flowers, with their petals deeply cut into four lobes, present an unforgettable and somewhat ragged appearance. The slender, reddish stems are about 60 centimetres tall, with lance-shaped leaves arranged in pairs along their length. Ragged-robin is still called by its country name of *cuckoo flower* in some places, because its appearance coincides with the calling of the *cuckoo*; its Latin name, *flos-cuculi*, also reflects this fact. Folklore associates plants with 'robin' in their name with goblins and evil, and it is still considered unlucky to pick this plant and place it indoors.

The pond here at Heyshott Green is a delightful place, in its setting of a natural and unspoilt village green with a backdrop of wooded hills. Many reeds and aquatic plants grow around its margins, one interesting one being *brooklime*. This plant has very attractive bright blue flowers and its oval, fleshy leaves growing in pairs along the stem used to be eaten in salads, having a similar taste to water-cress but much stronger. The *British Herbal* of 1756 suggested that 'a large quantity of this herb (brooklime) put into beer, while brewing, gives it the virtues of an anti-scorbutic (anti-scurvy) and sweetener of the blood in a very happy manner'.

You will not fail to notice the pair of *Canada geese* that have made their home on the pond. In early summer they are quite likely to be swimming around with their offspring, seeking food from passing visitors. The Canada goose is the largest goose in Europe at just over a metre in length. It is grey-brown, with a black head and long black neck, which contrasts with its whitish breast. It also has a broad white patch from the throat up to the cheek. This species was introduced as an ornamental wildfowl from North America in the seventeenth century. Further introductions were made to private lakes and waters in the eighteenth and nineteenth centuries, and by the 1930s they had spread across Britain. The Canada goose population has expanded rapidly and in East and West Sussex alone, there are now probably more than 150 breeding pairs. In the autumn and winter when their numbers are augmented by birds from neighbouring countries their numbers increase to 3,000 or more. The Canada goose makes its nest with locally gathered vegetable matter and is lined with down and breast feathers. The nest is in a hollow beside the water's edge on a pond or lake and usually on an island. The four to seven white eggs are laid in March and April and the grey goslings hatch out some 28 days later and soon take to the water where they are carefully guarded by their parents.

The Canada goose population is now so great that it is a nuisance to farmers, because as well as feeding on grass and aquatic vegetation, they also eat cereal and grain. They also do a great deal of damage by fouling footpaths and destroying waterside vegetation. Recently, certain control measures have been introduced on some sites to limit their numbers.

A good place to climb up the hills again is by way of the public footpath that ascends steeply up from nearby Graffham. As you pass the church at the end of the lane, a cock *chaffinch* gives a series of anxious chirps, 'wheet-wheet-wheet', from the branch of a tree, a quite different sound to that of its bright and cheerful song. This song is a brief vigorous cascade of about a dozen notes terminating in a flourish 'choo-ee-o', just like a cricketer when he makes his run up to bowl. One of our most numerous and widespread birds, the chaffinch, is easily recognisable with its bold white shoulder patches and in flight, by its white outer tail feathers. Seen close at hand, the male is a truly handsome bird with a lovely pale pink breast, warm pinkish-brown back, set off with a greenish rump and a slate-blue crown and nape. The female is less well adorned, with the same white markings, but generally of a dull grey colour and with none of the smartness of her mate. As with most birds in winter the male is much less bright, but still distinctive. In early spring, chaffinches pair up and it is at this time that the cock will sit on the top of a tree and sing, proclaiming his right over his chosen territory. The female bird does most of the work in building the nest in the branches of a tree or in a hedge, about 1–4 metres off the ground. The compact, cup-shaped nest, is made of moss, wool and grass, all beautifully felted together with cobwebs and lined with feathers and down. For concealment, the nest is finished off with a decoration of grey lichens on the outside. The eggs are laid in April with second broods starting as late as June. The female alone incubates the eggs, but the male is a very considerate partner, feeding her when on the nest, and bringing a plentiful supply of caterpillars to thrust down the ever-ready open beaks of the young.

As autumn approaches and the leaves begin to fall, so chaffinches start to congregate in small flocks, generally in single-sex groupings. In some winters, flocks of chaffinches migrate from mainland Europe to join resident birds feeding on weed seeds and insects in the fields.

On the flint walls near the church grows the *ivy-leaved toadflax* with its lilac flowers, each adorned with yellow and white lips. This plant was introduced to Britain in the seventeenth century as a rock garden plant and has now spread throughout the country, establishing itself on walls, banks and pavements, where it can be seen flowering from April onwards to October or even later. The seed capsules are carried on long stalks which bend away from the light so that the seeds are deposited in dark crevices. Once the plant is established it spreads by long-rooting runners which trail across the wall up to 70 centimetres long. This plant is particularly resistant to drought having roots that penetrate deeply into its chosen habitat.

As you climb up the steep, thickly wooded escarpment of Graffham Down there are wildflowers of all kinds growing on the chalky banks. In places under tall beeches large flints protrude through the mossy ground, *common dog violets* show their blue flowers from April to June together with *bugle* and *common spotted-orchids*. *Ploughman's-spikenard* is common here, and there are occasional clumps of *spurge laurel* to be found, easily seen with their shiny green leaves.

On top of the escarpment at Graffham Down you will see what looks like the remains of ancient earthworks. These are almost certainly 'pillow mounds' which were specially constructed of earth, wood and flint on the South Downs for the breeding of *rabbits*. This animal was introduced into Britain for meat by the Normans and was regarded as an important source of food. The 'pillow mounds' or rabbit warrens were looked after by a warrener from a nearby hut and were an important source of cash for the landowner in the Middle Ages. The mounds here above Graffham were some of the largest ever built, being 30 metres long, 5 metres wide at the base, and over a metre high. At one time they would have been honeycombed with tunnels and specially built nesting places. Deadfalls were constructed within them where the animals could be easily caught and then culled. Of course, over the years many rabbits escaped into the fields and survived in a feral state to become the familiar wild animal we know today. They bred quickly and by the 1950s they had become such a pest to farmers that myxomatosis, a highly infectious viral disease of rabbits, was introduced to control them. This was effective and by the 1960s their population had declined sharply. However, in the 1980s and 1990s the rabbit developed an immunity to the disease and its numbers are increasing rapidly with the consequent damage, not only to agricultural crops, but also to the ecology, and particularly to the flora of the South Downs.

In winter the paths up the side of the escarpment here are hard going with deep ruts filled with chalky water and with dark orange-brown *beech* leaves scattered everywhere and an occasional spattering of silver from those of the *whitebeam*. Below you as you climb, a thin plume of pale whitish blue smoke gently rises in the still air from a distant bonfire and a *blue tit* calls 'zee-zee-zee' high up in the bare branches. As you reach the top, you notice the large number of old *yew* trees growing, and quickly realise that the whole area here is thickly wooded, your views of the surrounding countryside being very restricted, particularly by the many conifers. However, local conservationists have kept a small area of open grassland for orchids and other downland species and have preserved ancient tumuli which show up as bell-shaped mounds beside the South Downs Way.

Here at the top of Graffham Down there are many pathways which lead off in several directions and these were once important links between the Manor Houses at East Dean and Singleton, with the woodland pastures of the Weald. In Saxon times these north-south tracks acted as droving roads between the settlements on the Downs and coastal plain and the deep Weald. People with their cattle and pigs travelled along them and crossed the escarpment by way of the inclined bostals mentioned in Chapter 5. So these track-ways are full of the history of South Downs people and are now used not so much as a necessary means of travel, but as a way by which people can enjoy the countryside and take pleasing exercise.

Follow the path that leads down through mixed woodland and open glades to Eastdean Wood. Along the way there is interest for the naturalist at all times of the year. In the spring, *cowslips*, *prim-*

roses and *bluebells* give colour to the woods and pathways and in summer the area has a wider variety of other wildflowers such as *nettled-leaved bellflower, ploughman's-spikenard* and *foxglove*. In places on shady banks the attractive *wood spurge* is common with its evergreen-looking oval leaves and bright golden, yellowish-green flowers. These flowers are borne in separate male and female clusters at the end of shrubby stems about 60 centimetres tall. The whole plant is covered in fine short hairs and often has a tinge of red on its stems and leaves.

As autumn draws near and the rains come, so the woodland floor takes on the smell of damp decay, a mustiness that suggests mould and the death of a place that was once full of life and vitality. However, this is just a passing phase of nature and as the beech leaves start to turn yellow, fungi of many varieties begin to form on old wood and on decaying vegetation. *Common ink caps* with crowded gills which rapidly turn black, and drip inky fluid as their spores ripen, grow in clusters on areas of rotting wood. These can be eaten but cause nausea and palpitations of the heart if consumed with alcohol. Good black drawing ink was once made from this fungi by extracting the liquid from it, and then mixing it with water and cloves. On the stumps of old trees, dense clusters of sulphur-yellow fungi aptly called *sulphur tuft*, can be found in many places, but like the ink cap, they can sometimes be found at other times of the year and not just in the autumn.

Beeches are the commonest trees in Eastdean Wood and in places they can be seen growing in neat rows with their trunks tall and straight and their side branches neatly trimmed away. These were clearly planted by man and this is very much a woodland which is carefully managed on a commercial basis, yet at the same time providing a wide variety of habitat for wildlife. In places there are plantations of *Douglas fir* which was first discovered in 1791 on the west coast of Vancouver Island in Canada. Its speed of growth has made it a major species in the British forestry industry, and its strong wood provides excellent constructional, flooring and joinery timber, as well as having many other uses which include fencing, paper pulp and telegraph poles.

In the depths of winter these vast acres of woodland take on a bleak and sombre appearance but there is life all around. A *dunnock* only some 14 centimetres long, with its grey-brown crown and brown back with black stripes, sings its 'wee, sissy-weeso' song, and a *wren* sounds its harsh churring 'titz, titz, titz', for this little brown bird delights to play the sentinel. When the winter sunshine gives just a touch of warmth, one may see *gnats* dancing in the amber light, their rapidly vibrating wings reflecting the transient gleam of rainbow colours. *Rabbits* feed at the edge of the woods in the early morning and as night falls, the dog *fox* can be heard barking as he travels through the shadows looking for a mate.

Much, much more could be written on this delightful Eastdean Wood and its surrounding forests of Charlton and Singleton, but there is so much yet to explore. Take the narrow, sunken path that leads to Newhouse Farm and on to the village of East Dean itself. As the path nears the farm and the tarmacadam road, count the different species of shrubs or trees that grow in a 25 metre stretch of hedgerow. The age of the hedge can be estimated easily for each different species accounts for approximately 100 years of age. There are at least ten different species here including *hawthorn, blackthorn, ash, spindle, dogwood, hazel, field maple, dog rose, elder* and *sycamore* giving an age of 1,000 years or so.

The village of East Dean is soon reached with its twelfth-century church, an impressive simple cruciform building with a massive tower and great oak beams. There is a lovely row of flint-built cottages where, unusually, even the corners are of extra large flints, beautifully worked in and meticulously placed.

Beside the road just a hundred metres or so to the west of Charlton there is a style and a path that leads up to Levin Down, another nature reserve. This reserve supports one of the last remnants of chalk grassland owned by the Goodwood Estate and has been managed since 1981 by the Sussex Wildlife Trust. Much of the surrounding area is now either forested or is cultivated, and the reserve forms a refuge for wildlife particularly plants and butterflies.

For some years after the Second World War Levin Down's steep south and east-facing slope was neglected and this caused scrub of hawthorn, blackthorn, dogwood, and bramble to spread over the area leading to a loss of open habitat. Careful management by the Trust has now created a wonderful site for chalk-loving plants such as *round-headed rampion, autumn lady's tresses orchid, clustered bellflower, marjoram, kidney vetch* and *horseshoe vetch* to name but a few. There is also an extensive colony of *juniper* and during the spring and summer you may see many of the 40 species of butterfly that occur here such as the *chalk-hill blue, painted lady, brimstone* and *orange tip.*

The *brimstone* is one of the first butterflies to be seen each year, sometimes as early as February. This sulphur-coloured butterfly is a strong flyer and roams many kilometres along hedgerows, around wood margins and in thickets and shrubby areas. It is one of the chief pollinators of *primroses* and has a long tongue to reach the nectar at the base of the flowers. It often stays for long periods at one flower and always remains with its wings shut, their yellow undersides blending in well with the vegetation. This camouflage, together with its 'hooked' wing shape, helps it to hibernate successfully through the winter as a butterfly, amongst ivy leaves.

The *orange tip* is another butterfly of early spring, the male of the species being easily recognised with its orange wing-tips. The female lacks the orange and has just the black on its wing-tips looking more like just another *green-veined* or *small white* butterfly. The orange tip very seldom alights with its wings open because the beautiful dappled green and white coloration of its undersides provides perfect camouflage amongst flowers and other vegetation. While the male patrols his territory, the female flies in search of food plants on which to lay her eggs. The eggs are laid singly on the flowers of members of the cabbage family, such as *lady's-smock* and *garlic mustard*, are bright orange in colour and are quite easy to find. The bluish-green caterpillars hatch out after about 25 days and then feed on the flower buds of the host plant. They are known to be cannibalistic and will often eat any smaller orange-tip caterpillars they encounter. They have long forked hairs which secrete a sweet liquid upon which ants feed, but what the ant does in return is not known. Only one generation is produced each year and the adult butterflies only live for about 18 days.

At the top of Levin Down there is a good area of chalk heathland where, during the ice age, acidic wind-blown material was deposited on top of the chalk, allowing acid-loving plants to grow next to those that thrive on the chalk. These chalk heaths are uncommon, and this one on Levin Down is the largest in West Sussex and is next in importance to the 62-hectare nature reserve of Lullington Heath in East Sussex, which contains the biggest chalk heath in Britain. The top of Levin Down therefore has a curious mixture of *heather* together with typical chalk grassland plants like *pyramidal orchid*, *wild thyme* and *round-headed rampion*. In places here at the top and on the slope you will also come across the *common gromwell*.

The *pyramidal orchid*, with its pyramid-shaped, rosy-pink, closely packed flower-spikes borne at the top of 15–50 centimetre tall stems, is one of the prettiest and most attractive of our orchids. Flowering from June to August, it is common just about everywhere on the South Downs and is sometimes seen in large numbers particularly on chalky banks and steep slopes. The unspotted leaves are narrow lance-shaped and taper to a sharp point. This orchid has a reputation for quickly establishing itself on new ground and is one of the first of its kind to appear where scrub has been cleared.

Growing amongst the orchids here is the *round-headed rampion* with its small, dark blue flowers closely clustered at the end of long stems to form globe-shaped heads, 2–3 centimetres in diameter. The whole plant, which has basal lance-shaped leaves with long stalks and rounded teeth, is about 15–40 centimetres tall and blooms from July to September. In Britain it is most common on the chalk grasslands in Sussex and is known as 'The Pride of Sussex'.

A somewhat local plant, the *common gromwell* with its small greenish-yellow flowers and upright appearance, is quite common on the slopes of Levin Down. It has stalkless, pointed, and oval leaves, with prominent veins and is much branched and about 30–60 centimetres tall. The flowers bloom in June and July and the leaves and stems are covered in hairs.

As already mentioned, Levin Down has been well managed to improve and maintain a good habitat, particularly for plants and butterflies. This has been achieved by the hard work of volunteers who have cut back acres of encroaching scrub over the years. In addition, with the agreement and co-operation of the Goodwood Estate, *sheep* grazing has been carefully controlled so that they are on the reserve until the beginning of April and return after the summer, so ensuring that the outstanding flora is allowed to grow, flower and set seed.

Stand beside the clump of trees at the top of the big curved hill that makes up Levin Down and look to the south at the huge hill fort of ancient man, the Trundle, with its earthworks of banks and ditches each interrupted by causeways like natural drawbridges. These causewayed camps were once used as corrals to round up cattle. Now, the Trundle glistens with the reflected sunlight off hundreds of parked cars. Nearby, the great grandstand of Goodwood racecourse and the drone of light aircraft from its airstrip, emphasise how modern man has tried, not always success-fully, to blend in his pursuit of pleasure with the natural beauty of the surrounding countryside. The inevitable discordance cannot take away the fact that so many people get so much benefit and joy from Goodwood's attributes, and surely this cannot be wrong. Before setting off for our next destination, Halnaker Hill, view to the west the great forests near West Dean, the remote countryside around Marden and the great curved Downs around Kingley Vale; these will form the starting point of the last chapter.

Halnaker Hill and its well known windmill is for me a special place. I remember one autumn morning walking up the old Roman road of Stane Street where for a brief kilometre or so it departs from the main A285 road. This deep track once reverberated with the marching of Roman foot soldiers, their footwear pounding on flint pathways and their armour clanking, as they travelled between Chichester and London. Alongside this track I counted some 17 different species of tree and shrub in a short stretch of hedgerow giving an age of some 1700 years, not a bad estimate since we know it originated in the Roman period. Amongst the species was *spindle*, its orange-red fruits showing up brightly against an early morning sky, and *field maple*, with its swaying branches still covered with yellow leaves and catching the light of the low sun as it emerged from behind a cloud. The trackway slowly ascends the side of the hill but as you leave the Roman road and climb over a style, the magnificence of the windmill on top of Halnaker Hill soon comes into view. The site seems to have been popular for mills; one is recorded here as early as 1540. This present one was built to a simple design and remains one of the oldest in existence. It ceased working in 1900 and in 1913 high winds carried the sails away. However these were replaced in 1934 and more restoration was carried out in 1955 when it passed into the care of the West Sussex County Council. Now, although all the interior machinery has been removed, it is an attractive sight with large white sails set against the red of brick and tile, mottled here and there with patches of ageing grey mortar. On my visit, the white sweeps of the sails and the white rounded dome shone brightly in the transient sunlight and were highlighted by a back-drop of a mackerel sky, beneath which were low scudding grey-black cumulus clouds.

The view of the surrounding countryside from this high point is magnificent. The sea sparkles to the south as do nearby glasshouses which cover vast acres of cultivation. Chichester Cathedral is plainly visible, its lofty spire showing well above the distant horizon and giving a sense of the spiritual to this lonely place. A small patch of reflected light indicates the site of one of Chichester's many gravel pits. To the north are the great forests of the Goodwood Estate, exquisite in the colours of autumn, areas of *larch* showing up as orange-grey, *beech* trees a mixture of orange with shades of red, and dotted here and there, patches of *conifer* and *pine* stand out with contrasting shades of dark green. Nearby, and all around are fields, some brown with newly-turned soil, some green with crops. In the middle of a nearby field a solitary *holm oak* stands in lonely isolation, its evergreen leaves densely covering its spreading branches and forming a long shadow on the ground. This *evergreen oak*, as it is sometimes called, was introduced from the western Mediterranean some 400 years ago. It is a common tree in gardens and parks of seaside towns, where it is particularly useful because it gives shade and shelter and resists the ill effects of salt-laden winds and atmospheric pollution. It has acorns similar to the *English* or *common oak* but they are shorter and are more deeply enclosed in their cups. The holm oak grows to a height of 20 metres or more and is a fairly common tree on the downland landscape.

Just to the south of the windmill next to the approach path is a large fallow field which is covered with the remains of wildflowers like *coltsfoot, salad burnet, yarrow* and *common St John's-wort.* Some species are still in flower on this warm, late November day; *dark mullein, wild basil, bristly ox-tongue, blue fleabane, ragwort,* and *ploughman's-spikenard,* to name but a few.

This area of Halnaker Hill was once a place with wide open spaces of uncultivated and somewhat desolate downland. Then it would have been a suitable place for that enigmatic and evocative bird the *stone curlew* to breed. This shy and stealthy bird was a fairly common breeding species early in the last century, but large-scale conversion of the Downs to arable land led to the loss of its habitat. For a while it nested in fields of spring sown maize, kale or beet, but soon gave up the task because of disturbance.

The stone curlew, about 40 centimetres in length, has streaked sandy-brown and white plumage, and can be recognised by its round-headed appearance with large yellow eyes. It runs furtively with head low, and flies usually close to the ground with deliberate wing-beats and occasional long glides. When in flight, the two bold white bars on each wing are conspicuous. Its nest on the ground is just a shallow scrape, lined with pebbles and rabbit droppings in which the two eggs are laid in April and May. Night-time is when the stone curlew is most active, and its wailing call 'coo-ree, coo-ree' is haunting and somewhat sinister when heard on a lonely heath in moonlight. There have been tentative plans to reintroduce this species to suitable large open areas of rough ground, particularly on established nature reserves. Let us hope that one day this bird will once again be seen and heard on the South Downs. In any event, you may be just lucky enough to catch a glimpse of it during its spring or autumn migration to and from Africa, the best places to see it being on the coastal beaches of Pagham and Rye or in the area of Beachy Head.

Before saying goodbye to the windmill, take a close look at the ageing brick-work. Growing here are tiny *harts tongue* ferns and *stinging nettles*, their deep roots finding sufficient food from the mortar and vegetative debris between the bricks. On the surface of the bricks grow several species of *lichen*, some very obvious with their conspicuous orange colour, others being less so, their greyish-green hue blending in well with their surroundings. There are 1,355 British species of lichen which are simple plants, each consisting of a fungus in association with an alga. The alga cells contain green chlorophyll and manufacture sugars and other compounds to sustain the plant, while the fungus provides shelter for them and prevents them from drying out. Lichen do not have roots, but absorb water and gases through their upper surfaces and are there-fore sensitive to atmospheric pollution. For this reason, they are rarely found in large towns and cities and prefer to establish them-selves on rocks, walls, tree-trunks, gravestones and roofs, particu-larly in the wetter west side of Britain.

The origin of the name Halnaker has puzzled many for a long time. The popular explanation is that it refers to a lack of cultiva-tion on one side of the nearby village of the same name, in other words half an acre or even half-naked. At various times it has been spelt in different ways – Helnache, Halnac, Halfnakede and Holnaker. Hilaire Belloc, early in the last century, captured so well the mood of decay and desolation of this mill, and the hill on which it stands, in his poem 'Ha'naker Hill':

> Sally is gone that was so kindly
> Sally is gone from Ha'nacker Hill.
> And the Briar grows ever since then so blindly
> And ever since then the clapper is still,
> And the sweeps have fallen from Ha'nacker Mill.
>
> Ha'nacker Hill is in Desolation:
> Ruin a-top and a field unploughed.
> And Spirits that call on a fallen nation
> Spirits that loved her calling aloud:
> Spirits abroad in a windy cloud.
>
> Spirits that call and no one answers;
> Ha'nacker's down and England's done.
> Wind and Thistle for pipe and dancers

And never a ploughman under the Sun
Never a ploughman. Never a one.

As you descend the hill early in the morning and before the many sightseers arrive, you leave behind the silent solitude of the mill which once was so busy with the making of flour, and in the Second World War was used as an observation post and as a navigational aid for aircraft. History is all around; the ruins of Halnaker House situated just 2 kilometres to the south-west was first built in the twelfth century and later rebuilt by the ninth Lord De La Warr, who, in 1520, attended Henry VIII at the Field of the Cloth of Gold and later entertained him at Halnaker.

Just to the east of the old Roman road near the bottom of the hill is a low chalk pit whose sides have long been worn smooth by motorcycles and bicycles. However, there are still good areas where chalk-loving plants can grow and here you will find plants such as *carline thistle, viper's-bugloss, common century* and the uncommon and interesting *white horehound.*

White horehound grows in clumps almost 50 centimetres tall and has rounded oval, blunt-toothed, wrinkled leaves which are green above and white beneath. The whole plant is covered with cottony hairs, giving it a hoary appearance. The flowers, unlike most of the mint family, to which it belongs, are white in colour and bloom from June to October. White horehound still enjoys considerable reputation as a cure for coughs and other lung troubles, a use that goes back at least 2,000 years. It has long been cultivated in the garden and was widely used by country people who took it in the form of a wholesome tea. Lozenges containing an extract from its leaves are still sold in shops today. These many uses for the plant are probably the reason why it is so uncommon in the wild.

White horehound should not be confused with the *black horehound* which does not have any medicinal properties and is a much more common species, flowering from June to September beside hedgerows, roadsides and on waste places. Black horehound has an unpleasant smell, hence its other name 'stinking Roger', and has pale reddish purple but sometimes white flowers. Although it has grey green hairy foliage it does not have the woolly hoariness of the white horehound.

Around the edges of fields that lie to the north of the chalk pit, you will often find in late summer and autumn large numbers of the tall (6–10 centimetres), greyish-brown fungi called *volvariella.* Their caps, 5–10 centimetres across, are sticky when wet, and the white stems taper upwards from the base which is enclosed in a whitish bag-like vulva. This fungi is edible but care should be taken in identification to avoid confusion with some other species which are deadly poisonous.

As you retrace your steps down to the bottom of Halnaker Hill, the sea catches your eye as it gleams in the sun some 18 kilometres to the south. This chapter closes with a visit to the important nature reserves at Pagham Harbour and Chichester Harbour.

Pagham Harbour Nature Reserve is managed by the West Sussex County Council and consists of a large lagoon protected by a one kilometre-long shingle spit with a narrow tidal entrance. The lagoon itself comprises mudflats, pools, salt-marsh and grazing marsh, and attracts many wildfowl such as *smew* in winter. There are many places from which to view the reserve, including the visitor centre just south of Sidlesham, but my favourite approach is by the path that leads south from the parish church of Church Norton. This takes you to the beach and it is worth looking closely at nearby bushes and scrub for autumn and winter migrants like the *firecrest*, with its sharp black and white eyestripe and golden crest. You may hear its shrill, high call 'zit-zit-zit' before you see it. *Short-eared owls* are often seen flying to and fro over nearby fields on the look-out for their favourite food, the *meadow vole*.

As you near the sea, you hear the soft hissing of the pebbles as they are turned over and over, and up and down the steeply shelving beach by the waves. In a storm the noise becomes a roar, as huge unrelenting waves pound the shore, and mists of white sea spray are carried far inland over the lagoon. The shore itself is littered with *oyster* shells and the white boat-shaped, chalky bones of dead *cuttlefish*. Each of these bones is a modified internal shell made up of five bony plates enclosing gas-filled chambers; they act as a buoyancy organ for the cuttlefish which is a close relative of the *octopus* and is common around the southern shores of Britain.

In the late spring and early summer the flora of the beach is at its best; *biting stonecrop* forms a mat of bright yellow flowers over the pebbles and huge clumps of *seakale*, their white flowers and fleshy green leaves, strikingly visible from a good distance away. The *yellow horned-poppy*, too, is easily seen, with its branching stems spreading in all directions and supporting large golden-yellow four-petalled flowers. These flower petals soon fall off after they are fully expanded, to be replaced by long (up to 30 centi-metres), sickle-shaped seed pods. Other plants you will see here are the *sea campion, red valerian* and the rare *childing pink* which we found at Shoreham Beach and which has another outpost here at Pagham. In places you will notice large leaves growing out of the

181

shingle, shaped not unlike an animal's hoof and covered with white cottony down. This is the *coltsfoot* plant whose yellow flowers first appeared in February, now long since gone and replaced by the leaves. When dried these leaves were once used as tinder to light fires, and they were also smoked as 'tobacco' to give relief from coughs. The Romans prized coltsfoot and used it to relieve asthma and bronchitis. Even today the rhizomes of this plant are boiled in sugar, and the resultant 'coltsfoot rock' is used to soothe sore throats. In Paris coltsfoot flowers were often painted on the door-posts of apothecaries' shops as a sign. Its Latin name, *tussilago*, derives from the word *tussis*, meaning a cough, and indeed, ironically, the leaves are still smoked as a substitute for tobacco.

The shingle spit itself, the salt marsh, and the islands in the lagoon, are nesting places for many birds including the *oyster-catcher*, *redshank*, *ringed plover* and *terns* and for this reason it is strictly forbidden to trespass onto it during the breeding season. However, you will be able to view the birds through binoculars from beside the warden's caravan. The haunting sound of the redshank calling across the marshes on a still day and the piping call of the ringed plover are just some of the delights of the place. Another is to sit on the beach in summer and watch a *common tern* hover over the sea on long slender wings before diving headlong into the water for sand-eels and small fish. This grey and white bird with a long black-tipped red bill, and black crown, migrates to and from southern Africa and arrives in early May, departing in September. It nests only rarely at Pagham Harbour and the ones you see are probably those that lay their eggs on the specially prepared rafts at Chichester Gravel Pits.

One interesting insect that occurs on the salt marsh at Pagham Harbour, is the *short-winged cone-head*, a green bush cricket that is found only in southern and eastern England. About 18 millimetres long, it is distinguished from other bush crickets by having a brown stripe down the back and a pale brown underside to its abdomen.

Before leaving this area of the shingle spit it is worth walking a little way along the beach to look at some small pools and reed beds called the 'Severals'. Here is another place which is good for migrant birds, and even rare vagrants like the *purple heron* have been seen here.

This place is now very different from when I knew it just after the Second World War. Then it was a wasteland covered with sunken pillboxes and huge square concrete blocks constructed as a defence system against tanks. Other defences marred the landscape then; scaffold-like structures, rusted a dark brown, stood as immense skeletons looking seawards, a reminder of man's foolishness. Luckily, all these trappings of war never had to be stand the test of a seaborne invasion. Now, the defences are gone, peace and tranquillity have returned and man can look on nature here, not with memories of war, but with what the peace has given back; a place of beauty hopefully to be enjoyed forever.

A short distance away lies Chichester Harbour, an immense expanse of sheltered water bordered with tidal mudflats, and salt marsh. This place is of national importance as a wintering ground for thousands of ducks, geese and waders. At low tide, wide areas of mud are uncovered and birds feed on the exposed marine animals. As the tide rises, the birds move closer to the shore and good views of them can be obtained from the perimeter footpaths, particularly at Hayling Island and East Head, West Wittering. At high tide many of the birds fly on to surrounding fields and pastures so it is worth consulting a tide timetable.

Many species of gulls and terns breed here and it is a good place to see the *little tern*. Fifty years ago there were many relatively undisturbed sandy and shingle beaches on the southern coast of Sussex and Hampshire where the little tern could breed successfully. Now, it nests only on nature reserves where it is given the necessary protection, not only from human disturbance, but also to some extent from predation by *foxes*, *crows*, *gulls*, *rats* and other egg thieves. The little tern is a migrant from Africa arriving in April and May and departing in late August and September. It is easily distinguished from other terns by its small size (24 centimetres), black-tipped yellow bill, black crown (in summer), white forehead, and yellow legs. It nests in colonies just above the high-water mark on sand spits, shingle shores and on islands. Its two or three greenish-grey eggs, blotched with grey and brown, are laid in a slight hollow in the shingle in May and June. The nest hollow is made simply and rapidly by the adult bird falling forward on its breast and scratching backwards with its feet and then pirouetting in a perfect circle. Incubation duties are carried out by both sexes and the young hatch out in about three weeks. The young are fed sand eels, their principal food, and leave the nest after about two or three days. At this stage they are very vulnerable to the cold, and to the dangers of being drowned by rising water levels in a storm surge. Until they can fly, about 28 days after hatching, they are also vulnerable to the attention of predators which include *owls* and *hawks*. Adults will often attack intruders and you may well see them chase away *crows* or *gulls* by diving *en masse* at them, at the same time uttering a high rasping 'kree-ik' and a rapid chattering 'kirri-kirri-kirri'. In the nineteenth century, terns, particularly little terns, were killed in large numbers for the millinery trade and they were almost exterminated along the eastern coast of the USA. Now, of course, terns are a protected species just about everywhere; let us hope that the small numbers of little tern that breed on our southern coasts will remain stable, or perhaps increase in the years to come. Much pleasure will result, as people see their splendid elegance as they flutter in the sun with fast wing-beats and dive headlong into the sea with a small splash; a just reward for the hard work of nature conservation.

In the late summer many of the drier areas of the salt marsh around Chichester Harbour are covered with a colourful display of plants such as *common sea-lavender* and *golden samphire*. This latter plant is a local species belonging to the daisy family. It has narrow, thick, fleshy leaves which in the flowering season, July to September, are dense at the top of the stem. The stem is about 15–45 centimetres high and the golden yellow flowerheads, about 2 centimetres across, appear in irregular umbel-like clusters at the ends of short branches. What an attractive sight this plant makes as it blends with the colours of yellow, orange, pink and red reflected off the calm waters of the harbour from a setting sun disappearing below an autumnal sky. Seemingly far off on the horizon, the low line of hills, dark with their covering of trees, reminds us that much of the South Downs between Chichester and St Catherine's Hill still awaits exploration in the final chapter.

Levin Down

Halnacker Windmill

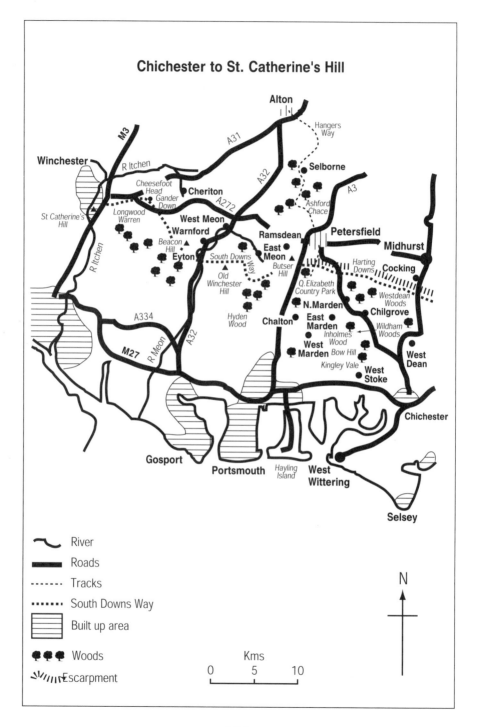

Chichester to St. Catherine's Hill

Alton

Hangers Way

M3

A31

A32

Winchester

R Itchen

Cheesefoot Head

Cheriton

A272

Selborne

A3

Gander Down

Longwood Warren

St Catherine's Hill

West Meon

Warnford

Ashford Chace

Ramsdean

Petersfield

Midhurst

Beacon Hill

East Meon

Harting Downs

Cocking

Eyton

South Downs Way

Butser Hill

Old Winchester Hill

Q. Elizabeth Country Park

Westdean Woods

R Itchen

A334

Hyden Wood

Chalton

N.Marden

Chilgrove

M27

R Meon

A32

East Marden

Inholmes Wood

Wildham Woods

West Marden

Bow Hill

West Dean

Kingley Vale

West Stoke

Chichester

Gosport

Portsmouth

Hayling Island

West Wittering

Selsey

~~~ River

━━ Roads

------ Tracks

••••••• South Downs Way

Built up area

Woods

Escarpment

Kms
0   5   10

N

186

8

CHICHESTER TO WINCHESTER

Butser Hill

In complete contrast to the placid blue waters of Chichester Harbour, Kingley Vale is a beautiful nature reserve, which, in 1952, was the first to be established in Britain and contains the largest yew forest in western Europe. Managed by English Nature, this outstanding reserve, not only supports a diversity of wildlife habitats, from chalk heath to dense woodland, but also instils in one a deep sense of history.

The best approach to this reserve is from the unclassified road at West Stoke, a village some 5 kilometres to the north-west of Chichester. A good car park has been provided at the village with clear signposts to Kingley Vale itself. A good time to see the reserve at its best is in June; even as you start off to walk from the car park along the track, you will see the purplish-red flowers of the *hedgerow crane's-bill*, some 50 centimetres tall, growing amongst the grasses. This plant is distinguished from the other species of crane's-bill, by its very hairy, rounded leaves which are cut into five to nine lobes. After about 15 minutes' walking, you will come to the main entrance where there is a field centre with a good permanent display and information about the reserve.

The largest of the *yew* trees grow at the foot of the valley and it is believed that some of them may have been established here in pre-Christian times. Many of them are dramatically gnarled into weird and fantastic shapes and, although aesthetically appealing, very little grows beneath them. In some places the great limbs of the trees themselves, contorted by time and centuries of storms, have broken and fallen to the ground and here new root systems have developed giving the appearance of huge, motionless serpents. At twilight, sometime after the sun has disappeared below the high hills, this place amongst the yews is one of haunting shadows, where with little imagination unquiet spirits of long-forgotten people seem to be all around. There are many local legends of witchcraft and hauntings so, clearly, this is not the time and place for those with a timid disposition.

Not only do the yews grow here, but many other trees and shrubs as well, including *beech, blackthorn, hawthorn, dogwood, holly* and *privet*. The berries and seeds of these provide an autumn feast for winter visitors like the *redwing* and *fieldfare*. Both these species of thrush migrate southwards from their breeding grounds in the forests of Scandinavia to escape the cold, harsh northern winters and to find food. The redwing can be recognised by its eye-stripe, reddish flanks, and when in flight, by the chestnut colouring under its wings. The fieldfare is larger, has a grey head and rump, and a rusty-coloured back. From late September onwards these birds can be seen in mixed groups seeking food in fields and hedge-rows, and they are particularly fond of fallen fruit in back gardens and orchards. Often one bird acts as a lookout, sitting as high as possible on a post, bush or tree, and when danger threatens, gives a harsh chattering alarm call which sets all the birds to flight. Redwings begin to leave for the north again in late March, slightly before the fieldfare, but both have gone by May.

The *song thrush* is another bird you may see in amongst the trees and bushes, especially in the autumn and winter when they move away from their breeding locations and form loose flocks on the Downs seeking berries, snails and worms. The song thrush is a well-known bird of our gardens, but I wonder how familiar it will be to future generations, for since the 1970s it has been on a steady decline in most parts of Britain. The reason for this is unknown, perhaps the widespread use of slug pellets has contaminated the food chain through one of its main sources of nourishment, snails; or perhaps its decline is due to a change in farming practices, or even the increase in one of its main predators, the *sparrowhawk*. The song thrush, unlike its relative the *blackbird*, does not often breed on the open downland even where there are plenty of suitable nest sites; it prefers to nest where there is good cover in woods, hedgerows and sheltered valleys. In early spring, the female makes its nest of grass lined with mud, about 2 metres up in an evergreen bush or hedge where it is well hidden. The nest is easily distinguished from that of the blackbird which is similar but lined with grass. The song thrush lays her three to six attractive pale blue eggs spotted with black in March or April, with repeat clutches on to as late as August. The cock bird is very attentive and feeds his mate while she is on the nest and helps with feeding the young. He is also a diligent housekeeper, removing hatched egg shells and carrying away excrement of the young. This thrush, as its name suggests, has a beautiful song, sung in most months of the year from a song perch, where the bird often sits with its head cocked to one side. The song can be easily recognised by the fact that each loud, but short and clear phrase, is repeated two to four times.

Song thrushes are particularly susceptible to cold weather and as a result they are migratory, always travelling towards the warmth. If the winter is harsh in northern Britain they will move to the south and west, even to fly on as far as Spain and Portugal. If it is very cold on the other side of the Channel, then continental birds will visit us to take advantage of the warm Gulf Stream which keeps our southern and western coasts relatively mild. The song thrush with its exquisite voice is one of the delights of our garden and gives much pleasure to many, and especially enriches the lives of those who live in suburbia, surrounded as they are by houses and traffic. Robert Browning clearly had these thoughts of this bird in mind when he wrote in his poem 'Home thoughts from Abroad':

> Hark, where my blossomed pear-tree in the hedge
> Leans to the field and scatters on the clover
> Blossoms and dewdrops - at the bent spray's edge –
> That's the wise thrush; he sings each song twice over,
> Lest you should think he never could recapture
> The first fine careless rapture.

You may also see along the edges of the wooded areas of Kingley Vale and within easy reach of fields and meadows, a larger version of the song thrush, the *mistle thrush*. It is also known as the 'stormcock' because it is almost the only bird that sings on a cold January day, even in a storm, and then it can be seen at the top of a leafless tree throwing out its loud, flute-like notes over a desolate countryside.

As you walk through the reserve the 'laughing' call of the *green woodpecker* reminds you that there are many different species of bird breeding in the area; these include the *sparrowhawk, tawny owl*, and many warblers such as the *nightingale, blackcap, willow warbler* and *whitethroat*. In the autumn you may be lucky to see migrant birds moving south such as the *ring ouzel, osprey, black kite* and *marsh harrier*.

Mammals are well represented on the reserve where *rabbits* and herds of *roe deer* and *fallow deer* help to restrict the invasion of unwelcome tall grasses and scrub onto the open grasslands. *Stoats, weasels, badgers*, and *dormice* are also resident, but of these the only ones you are likely to see during the day and then with just a fleeting glance, will be the stoat and the weasel as they dash across your path.

In the eighteenth century a sheep fair was held annually in the valley on the 2nd June. At that time until the early 1950s there was little scrub or coarse vegetation and the yew grove adjoined directly onto close-cropped turf. When the days of the shepherd and his wandering flock came to an end and myxomatosis decimated the rabbit population, so pernicious scrub spread over this rich habitat of downland turf. Those managing the reserve in the early days fought a constant battle to prevent total destruction of the grassland. This struggle continues today and the management has produced a fine balance between scrub, thick vegetation and short turf, thus providing a diverse habitat suitable for a variety of wildlife. In the areas of open grassland there are many varieties of chalk-loving plants such as the *common rock-rose, clustered bell-flower* and *eyebright*, and many species of orchid including the *bee, fragrant, frog, twayblade* and *common spotted*. Even an old dew pond has been restored to ensure that a source of water is available for the numerous animals and birds.

The nature of this habitat, with its diversity of plants, provides just the right conditions for butterflies and of the 58 species that breed in Britain, 39 have been recorded at Kingley Vale. You are likely to see many of those already mentioned, like the *chalk-hill blue, marbled white, meadow brown* and *speckled wood*. Others you may see here are the *wall brown* and the *silver-washed fritillary*.

The *wall brown* likes the sunshine and is often seen in high summer on the sides of downland tracks where it loves to spread its brightly-coloured wings on stones or bare places. This is a butterfly that rises early and retires late, and its rich orange brown colours, with black markings and white eye spots, are particularly striking when seen glowing in the setting sun of a summer's evening. The male of this butterfly is very territorial and will perch on prominent places and fly down aggressively to drive away intruding insects, not unlike the *speckled wood* butterfly. He will also spend much time patrolling paths and hedgerows looking for females, and to take nectar from various wild flowers. The eggs are laid in many types of grass, such as *tor grass* and *false brome*, on which the caterpillars feed. The early summer caterpillars develop in a month, before turning into chrysalids, producing adult butterflies in July and August. Late summer caterpillars hibernate over winter to pupate in April and produce adult butterflies that are on the wing in May and June. The life-span of each butterfly is about three weeks. In very hot years when there is plenty of sunshine the wall brown is seen in large numbers on the South Downs, particularly on the sides of dusty arable fields and roadsides where its food plants are characteristically found. If the summer is cold and wet the butterfly is rarely seen and, of course, one poor year leads to reduced numbers in the following years. The practice of cutting back the grass on roadside verges, together with agricultural improvements in reducing the areas of rough grass around field boundaries, have both contributed to the decline of this attractive butterfly in recent years; a sad reflection on nature conservation generally.

You may also see the *silver-washed fritillary*, a large colourful butterfly, on the woodland margins and in the clearings, when from June to late August it is very noticeable as it rests with wings open, on leaves in the sunshine, or as it partakes of nectar from blackberry or thistle flowers. The upper surface of its wings, which span 7 centimetres, are bright orange and are boldly marked by black spots and streaks, while the undersides of its hind-wings are a wash of pale silver and green. This butterfly is unique in Britain in laying its eggs on tree trunks and not on the food plant. The spiny caterpillars hatch out in two weeks, eat their eggshells, hibernate in winter and then descend to feed on *dog violet* plants in the spring. The adult butterflies live for about five weeks and are most likely to be seen from June to late August. This butterfly, although not that common, can be seen in many woods of the western Downs, particularly where there is plenty of dappled shade, and wide sunny tracks and clearings with good patches of dog violet.

Before leaving this wonderful reserve, just sit for a while amongst the flowers of a small chalk heath near the summit of Bow Hill. Not only is this one of the most picturesque places on the South Downs, with splendid views, but it is also an interesting archaeological site, possibly the best in West Sussex. Here can be found the site of a Roman temple, ancient flint mines that are now just wide depressions in the ground, and all around even older neolithic trackways. Here too, just outside the reserve at Goosehill Camp, are widely spaced concentric ramparts considered to be the remains of an Iron Age fortified cattle enclosure. Lynchets and prehistoric earthworks are in evidence in many places, as are four magnificent round barrows of the Bronze Age which are called the 'Devil's Humps'; as on Chanctonbury Hill it is said that anyone running around them will be rewarded by a sight of the Devil in person.

In about AD 900 it is believed that a great battle was fought in the area between the Saxons and Viking raiders, and the name Kingley Vale is thought to have been derived from the notion that their leaders (kings) were buried in the barrows. Sadly the facts belie the fantasy, for these burial mounds belong to the Bronze Age about 3,000 years ago.

A *wood pigeon's* repeated cooing sounds from a nearby bush and as the lowering sun shines on distant Pagham Harbour, it is time to say farewell to Kingley Vale. So with the great feeling of history all around, where memorials of vanished people lie desolate and half forgotten on the hillside, you descend through the perpetual, sombre twilight of the woodland and leave behind this nature reserve, so majestic in its diversity and splendour.

Our next destination is the secluded, some may say remote, area of woodland, south of East Marden. Here, Wildham Wood and Inholmes Wood, both Forestry Commission areas, provide a refuge from the busy world of human toil, for here are places where one can feel utterly lost and alone. These mixed woods, mainly of *beech*, cover just part of the southern portion of this sparsely populated area of the four Mardens (East, West, Up and North), little villages where flint walls and cottages are set amongst steep meadows, hanging woods and arable fields; so refreshing and seemingly untouched by the busy world around.

In the spring these lonely woodlands are a delight for the walker. I remember once, as I strolled along a path deeply covered with decaying beech leaves, my eyes being drawn to a patch of bright white on the woodland floor some distance away. As I got closer, I was thrilled to discover a small group of *greater butterfly* orchids, their creamy-white flowers almost gleaming in the shadows of the trees. A slight rustle amongst some thick holly bushes nearby caught my attention. I stood quite still; suddenly a graceful buck *roe deer* slipped quietly out from his concealment and through the undergrowth, to be followed by two fawns and the doe or female deer. The roe deer just 80 centimetres tall, is the smaller of the two native British deer, the other being the *red deer*, the largest British wild animal. The roe deer is common here in the western woods of the South Downs, where it is generally unobtrusive and solitary, except when in small family parties. Antlers are normally present only in the male and are short and erect; they do not spread and rarely reach a length of more than about 20 centimetres. The coat is a warm red in summer, changing to greyish white for the winter. I was lucky to be so close to the family on that May morning, as this deer is mainly nocturnal in its habits, generally lying up in thick cover during most of the day. Rutting occurs from mid-July to mid-August when the buck keeps other males away from his selected mate by uttering gruff barks. Courtship often involves the buck chasing the doe in circles, usually around a bush or tree, so forming a trampled ring on the woodland floor. This chase pairing, where the doe allows the buck to chase her round and round the ring, often lasts for some time until mating takes place. The fawns, frequently twins, are born from April to June. A word of warning is necessary here, for if you do happen to hear the unfriendly bark of a roe deer in the rutting season, keep away, for the buck is very pugnacious at this time and will attack any intruder who dares to get near the ring.

You may come across another deer, the *fallow deer*, which has also spread across the heavily wooded Downs in this area. Standing about a metre tall, this deer has a reddish-fawn coat in summer dappled with numerous white spots; this coat changing in October to its winter colour of uniformly greyish-brown without spots. The fallow deer has much larger and broader antlers than the roe deer, and it has a small tail; quite unlike the roe deer which has virtually no tail to speak of, just a small pimple. The fallow deer was introduced to Britain by the Romans from Asia Minor, who brought it as a decorative estate animal and as a beast to be hunted for meat.

Sadly, both the roe and fallow deer have found the South Downs woods so much to their liking that they have increased rapidly in numbers and are regarded as a considerable forestry pest because of the damage they do to trees; they now have to be kept in check by licensed marksmen.

These woods have much to offer the walker and naturalist, and in places, carpets of *bluebells* give pleasure in the springtime, with *primroses, herb robert* and *early purple orchids* adding splashes of colour to an already colourful scene. Butterflies of many species flit around in the open places and along the woodland edges; see how many different types you can count.

193

Autumn brings a mass of colour to these woods and the first chill winds of winter cause the leaves to begin their fall. Each slight breeze, and the air is filled with an unimaginable multitude of colours, green and yellow, gold and bronze, brown and crimson. The trees are soon denuded to the very last twig and then stand stark and naked, while the floor is seemingly covered with a carpet of golden brown; these leaves giving shelter and warmth to the many creatures which must endure the biting cold of the months to come. The winter wind moans through the trees and blows the leaves about on the ground; finally the wind abates leaving the wood to rest in peaceful slumber. Imagine the scene in the depths of winter after a few days of heavy snowfall. Stand alone amongst the softness and silence of the snow-covered woodland floor, where almost imperceptibly time itself seems to slow as the sun hovers in a bright blue, windless sky, and shines a fitful light through the tall, bare branches of the *beech*. A scene of such tranquillity and peace, seldom found anywhere else, must be savoured at leisure; so do not hurry; just enjoy it all. The tracks of many animals can be seen, leaving regular spaced holes with dark centres where dead leaves cover the soil on the ground. Take pity on these animals, whatever they may be, *fox*, *deer* or *badger*, for should the weather get colder and more snow falls, then their situation will become perilous from lack of food.

Edward Thomas described the scene so well in his poem 'Out in the Dark'.

> Out in the dark over the snow
> The fallow fawns invisible go
> With the fallow doe;
> And the winds blow
> Fast as the stars are slow.

West Dean woods are just some 4 kilometres to the north-east of Wildham Wood beyond the small village of Chilgrove. In early summer the roadside hedgerows are ablaze with *dog rose* and the wayside grasses are splashed with the colours of *red clover*, *lady's bedstraw*, *common spotted-orchids* and *oxeye daisies*, these latter particularly resplendent as their white and yellow flowers at the top of long stems sway back and forth, caught in the turbulence of passing traffic. Access to the Westdean Woods Nature Reserve is by permit only from the Sussex Wildlife Trust, who manage it. This reserve is just a tiny fragment of the woodland that once covered the South Downs before early man started to clear them.

Because the woods have been here a long time, the soil is very fertile and relatively deep, compared to that below chalk grassland. As with many ancient woodlands this is an area of *oak* and *hazel* which has been coppiced for many centuries to produce hurdles and fences from the hazel, and building timber and firewood from the oak. The act of coppicing where the trees are cut in winter near to ground level in a 7 to 20 year cycle, extends the life of each tree by several hundred years. In the following spring the cut stumps, known as stools, are stimulated to produce a new crop of young shoots to be cut in the next cycle. Coppicing was once an important part of the life of every village, supplying fuel and other woodland articles, but over the past 140 years as Britain modernised with coal and electricity, the demand for coppicing products declined and the industry gradually waned.

Here at Westdean Woods a wide range of wild flowers, insects, butterflies and birds have adapted to exploit the artificial conditions created by the coppicing that has been reintroduced since the site was created a reserve. During the first year, after cutting, the woodland floor is hot and dry, but after two or three years, there is shade, light and air; plants such as *primroses, violets, early purple orchids,* and many others begin to appear in large numbers, providing food for butterflies such as the *pearl-bordered fritillary.* As the coppice grows and becomes more dense, it gives good cover for birds such as the *blackcap, willow warbler* and *nightingale.* Eventually after seven years or so the canopy has regrown completely, excluding the sun and giving cooler and darker conditions. By careful management of this coppicing, the Sussex Wildlife Trust has provided a wonderful diversity of habitat for wildlife, the health of which is regularly monitored with the help of volunteers.

To the north of the reserve are two areas managed as superb habitats for *lichens* and *wild daffodils.* The Westdean Woods Reserve is well known for its large colony of wild daffodils; appearing from February to March, depending to some extent on the severity of the winter, these unmistakable harbingers of spring, with their deep yellow trumpet-like flowers, make a wonderful showing in the patches of damp grass beneath the trees. You will not fail to see them if you follow the public footpath that skirts the western boundary of the Reserve. Interestingly the word 'narcissus', describing the genus of bulbous plants that includes daffodils, gets its name from a boy in Greek mythology who was told he would be happy, as long as he never saw his face. Sadly he saw his reflection in a pool, fell in love with it, and pined away leaving a yellow flower on the spot where he died.

There are uncommon and interesting plants that grow in Westdean Woods such as the *toothwort, adder's tongue fern* and *greater* and *lesser butterfly-orchids*. In late March and April near the wild daffodils, you may spot the fungus-like flowers of tooth-wort, which have curious pinkish, leafless stems, fringed with fleshy scales. This extraordinary plant cannot be mistaken for any other and is a total parasite, lacking chlorophyll, and feeding entirely on its host, in this case the *hazel*.

Another curiosity in this area is the *adder's tongue fern* which, in May and June, puts out a single very unfernlike, pointed oval leaf, and a single green fertile spike up to as much as 15 centimetres tall. The spike carries the spores which germinate in July and August to form male and female organs which fuse with the aid of moisture, and develop into new young shoots the following spring. The adder's tongue fern is a plant of damp grassy places especially on calcareous soils.

The *lesser butterfly-orchid* is an uncommon orchid that grows in a few places on the coppiced woodland floor. This orchid is very similar to the greater butterfly-orchid described in Chapter 5, but there are differences in the structure and shape of the pollinia. In addition, the entrance to the long flower spur is narrow and appears closed, and the spur projects almost horizontally across the flower spike, not having the tendency to curve downwards in a semicircle, like the spur of the greater butterfly-orchid. The 15 to 20 cream-coloured flowers are held in a loose spike on the stem and appear in late May and June.

Follow the gently rising path northwards past the wild daffodils and oak woods and onto a vast open area with plantations of young *beech, larch,* and *Douglas fir* interspersed with *ash* and *silver birch*. Alongside the dry trackway grow *marjoram, common centaury, hemp-agrimony* and *wood sage,* and covering the low bushes are the trailing stems of *honeysuckle* and *wild clematis*. In places and quite unmistakable, are large patches of *coltsfoot* leaves with their cottony underparts and irregular wisps of 'cobweb' on their upper surface. The path gradually ascends until you are envel-oped by a tall plantation of *Norway spruce*, a fine forest tree, resis-tant to frost and grown for its tough wood, used for box-making, interior joinery, paper pulp and general carpentry. In June and July on the woodland edges, here, as well as many other similar sites on the South Downs, you will almost certainly see the *ringlet butterfly*.

The ringlet is a very dark velvety-brown butterfly with prominent white eyespots encircled by yellow, and with black rings on its under-wings. The female, which is larger and not so dark as the male, scatters her eggs during flight, so that they fall on one of the many common grasses that the young caterpillars will eat in autumn. The caterpillars hibernate in winter, occasionally feeding on mild nights, and then resume continual feeding from March to June. They form chrysalids in June which turn into adult butterflies after about two weeks. The ringlet may not appear very attractive at first sight, but watch it as it takes nectar from a bramble flower and as the sun's rays catch its conspicuous false eyes and the white, marginal fringes on its wings; then it becomes a delicate creature of beauty and enchantment. Sadly, like so many other butterflies it only has a lifespan of about two weeks.

To some extent, the dark, seemingly lifeless forest is oppressive, but beneath the trees are carpets of velvety emerald-green *moss* and soon the horizon ahead lightens and the path takes you across open cultivated fields just above the north escarpment. A line of huge gnarled *beech* trees stand as sentinels beside the track, relics perhaps of great forests long ago. The South Downs Way soon appears, traversing as it so often does, across the middle of these furrowed fields, and flanked either side by wire fencing, a cheerless trapping of modern farming. Ahead, just to the right, the trig point of Linch Down looks forlorn, surrounded as it is by upturned greyish-brown clods of newly ploughed earth. The views across wooded hills and arable fields to the distant Isle of Wight are quite superb. Follow the South Downs Way westwards and past four large bell barrows and a smaller one, all in a line; these burial mounds are known as the Devil's Jumps, one of which is over five metres high and the tallest of any in East or West Sussex. Soon Beacon Hill comes into sight and at 242 metres, it is one of the highest points on the South Downs. On the summit is a 40-acre, rectangular hill fort. Not much is known about this ancient site, but during the Napoleonic Wars there was a semaphore station here, one of a chain which sent messages from Portsmouth to Whitehall, this station being commemorated in Telegraph House, a kilometre to the south. From Beacon Hill, the South Downs Way descends and then climbs steeply to Harting Downs. Along this stretch of open grassland look out for the *Duke of Burgundy fritillary*, a butterfly that has never been common in Britain and is confined to scattered localities in woodland clearings and downland, mainly in southern England.

Late May and early June is the best time to see this small dark butterfly with orange spots, as it basks in the sun, low down on a leaf, and then makes a short flight to settle again nearby. It often flutters above the caterpillars' food plants – *primroses* and *cowslips*, on which the female lays her smooth, pale green eggs in May. These hatch out in two weeks and the brown, hairy caterpillars feed on the underside of the leaves until September, when they turn into chrysalids. They spend the winter in this form until the adult butterflies emerge the following May. The adults are very active flyers in their short lifespan of about 20 days, and they are attracted to the nectar of *wood spurge* and flowers that are blue in colour, but seldom visit other kinds of flower. A great deal of conservation work has been done on these western Downs in removing scrub to encourage primroses and cowslips and so allow this butterfly to spread and increase in numbers.

Harting Downs is a wonderful area of downland, smooth and bare, with a good chalk grassland flora and many butterflies. Here there is a local nature reserve, enclosed by National Trust property. On these lovely grassy slopes in June you will see many orchids such as the *twayblade, common spotted* and *fragrant*. Other plants are here to greet you, such as the *heath speedwell* with hairy creeping stems and lilac-coloured flowers, and the small, slender, and hairless *fairy flax* with its tiny white flowers. One of the many species of sedge grows amongst the grasses, the *glaucous sedge*, which has pale bluish-green leaves similar in colour to those of the carnation; hence its other name, *carnation grass*. The glaucous sedge is very common and grows just about everywhere on the chalk so you should have no difficulty in finding it. In the woods below Harting Downs you will come across many other interesting plants including *broad-leaved helleborine, bird's-nest orchid* and *Solomon's-seal*, this latter plant here being truly wild and not just a garden throw-out.

A dew-pond and some well positioned seats complete this idyllic spot; sit here and watch the world go by beneath you, as a *skylark* sings above you and a pair of *goldfinches* twitter amongst the thistle heads. Below you lie the old Saxon villages, Elsted and the Hartings – East, West and South. The tall church spire of South Harting is prominent as it soars above the surrounding buildings, trees and farmland. The slender spire is covered with copper which reflects many unusual and changing colours, especially when a summer sun sets low over distant hills.

The countryside below the escarpment here at Harting Downs is a good place to see the *spotted flycatcher*. This fairly common summer visitor and passage migrant is a bird of orchards, woodland edges, large gardens, churchyards and parkland. The spotted flycatcher is a small mouse-coloured bird, lighter below, with darker upper parts and dark streaks on its head and breast. Both sexes are similar in appearance and soon after arriving in this country in April from southern Africa, they pair up and busy themselves with nest building. The nest, made of roots, moss, hair, feathers and grass, is held together loosely with cobwebs and is built typically on the horizontal branches of fruit trees, trained against walls, or in holes in trees, or in crevices amongst ivy covered walls and trees. The four to six eggs are laid from May to July and there are often two broods during the year. Although somewhat inconspicuous, you are most likely to see this bird when it sits on a good look-out perch, such as a wall or fence post, to wait for passing insects. When a fly, butterfly, moth or small dragonfly flutters past, the flycatcher makes a quick dash after it and snaps it up, often with an audible click of its bill. Walter de la Mare described this so well when he wrote:

> Grey on grey post, this knife-winged little bird
> Swoops on its prey – prey neither seen nor heard!
> A click of bill; a flicker; and back again.
> Sighs Nature an alas? Or merely, Amen?

Once the flycatchers have hatched and fledged, they do not linger long around their home ground and by late August they begin their southward migration. Since the 1960s there has been a steady decline in the numbers of spotted flycatchers visiting our shores, and this is probably due to the drying up of the area just south of the Sahara, known as the Sahel region, through which the birds must pass on their migration.

Our route passes above the very steep escarpment above Leith Copse. Below are green cultivated fields and here and there are meadows dotted with flecks of white; sheep feeding contentedly in a peaceful countryside. In Saxon times the border between Sussex and Hampshire near Leith Copse was marked by the planting of *broad-leaved limes*. These have since propagated themselves and are now a towering grove of aristocratic trees. The wooded escarpment here is the place to look for *herb paris*, an attractive and extraordinary wild flower.

Herb paris is a local plant in Britain, growing in damp woodlands on rich calcareous soils, but for some reason it is very rare on the South Downs. It has a stout, creeping rootstock, and a rounded stem up to about 20 centimetres high, which bears a single whorl of four oval leaves held flat like a Maltese cross. From the centre of the leaves, the flower stalk rises, at the top of which, a solitary bloom appears in May and June. This flower is shaped like a star with four very narrow yellow-green petals and four wider green sepals topped by a crown of eight golden stamens. The flower has an unpleasant odour which attracts flies to effect pollination. A single, shiny black-hued berry soon forms on this, and then splits irregularly to discharge black seeds. The name 'paris' is derived from the Latin *par* meaning equal, an allusion to the symmetry of the four- and eight-part regularity of the species. In the dappled sunlight on the woodland floor, the leaves of herb paris mingle and merge with those of *dog's mercury* with which it is commonly associated. One of the delights for me when searching for this elusive plant, is to suddenly see the symmetrical pattern of four oval leaves, together with the subdued glimmer of yellow and gold, surrounding a berry of shiny black.

Our journey continues across the county boundary and on into the great Queen Elizabeth Country Park. This huge area of dramatic downland and beautiful woodland is part of the landscape of the South Downs and is an Area of Outstanding Natural Beauty (AONB) which, here in East Hampshire, extends from Chalton to just west of Warnford. The AONB was so designated on the 26th September 1962 and confers national recognition of the area's outstanding quality. It helps provide protection from unsuitable development through planning laws, and at the local level gives greater powers to the rural community and its planners to keep any necessary development in scale and sympathy with the area. The Country Park is managed by the Hampshire County Council in conjunction with the Forestry Commission and its 1,400 acres is a naturalist's paradise with 38 species of butterfly, and 12 species of wild orchids, to say nothing of the wide variety of other plants, insects, birds and animals that inhabit it. This is a place with so much to offer that a whole book could be written on it alone, so take time to pay a call at the visitors information centre situated just beside the busy A3 which divides the park into two. To the west is the open downland of Butser Hill and to the east is the huge and heavily-wooded area of mainly *beech*, dominated by the hills of Holt Down and War Down. It is hard to imagine that only about sixty years ago these hills, now clothed in forest, were open grassland with only scattered *yews* and *hawthorns* to break up the skyline. Then the whole area was a huge rabbit warren tended by a resident gamekeeper who was employed to maintain the rabbits for shooting, a highly popular sport at that time. An extensive network of trails and paths provide for walking, orienteering, horse riding and off-road cycling. Mountain-bikes can be hired, horse-riding lessons can be obtained and there are specific areas set aside for kite flying, model gliders, hang-gliding and paragliding. In addition, although not part of the Country Park, some 2 kilometres to the south is the Butser Ancient Farm, founded in 1972; this is a comprehensive reconstruction of an Iron Age farmstead dating back to about 300 BC, showing what it was really like to live more than two thousand years ago, and is an interesting place to visit.

Worth mentioning is the 34-kilometre and well-signposted Hangers Way walk, that takes you north from the Queen Elizabeth Country Park, through wonderful wooded countryside, to Alton. You will pass through many places of historical interest and experience the grandeur of great *beech* hangers with their rich ground flora. There are five nature reserves, and the one at Ashford Chace has a fascinating flora which includes *herb paris*, *bird's-nest orchid*, *narrow-leaved helleborine* and very rare plants such as *stinking hellebore* and *alternate-leaved golden-saxifrage*. Stinking hellebore is about 80 centimetres high, has yellowish green, bell-shaped flowers edged with purple, which bloom from January until April. It gets its name from the fact that if bruised it emits a most unpleasant smell. The golden saxifrage has two forms, the more common one, named the *opposite-leaved golden-saxifrage*, with circular shaped leaves, borne opposite each other on its many spreading rooting stems, and the much more local species which grows in damp places at Ashford Chace, the alternate-leaved golden-saxifrage. As its name suggests the leaves of this last species do not grow opposite each other on the usual single flowering stem, and are kidney-shaped. Both species have flattened heads which are made up of numerous small bright yellow-green flowers that appear in April and May.

An important part of the huge Queen Elizabeth Country Park is Butser Hill which on the 3rd September 1999 became Britain's 200th National Nature Reserve. I believe the unsurpassed beauty of this hill, with its open bare slopes of chalk grassland, is best seen from the north. Take the road that leads south from the village of Ramsdean, and then follow the track that starts from the underhill lane and transverses Ramsdean Down up to the summit trig point. As you walk this slope you will not fail to notice the short springy turf that lies beneath your feet. This typical downland habitat, so common on the eastern downs, is covered with wild flowers from spring to autumn. *Cowslips* are among the first to appear, to be followed by a multitude of others; orchids such as the *bee*, *fragrant* and *common spotted* are to be found, together with *round-headed rampion, hoary plantain* and *stemless thistle*; this last named, sometimes called the 'picnic thistle' on account of its sharp spines, guaranteed to penetrate the thickest of clothing when sitting down on the soft turf. In late summer the *autumn lady's-tresses* appear together with *carline thistle* and everywhere there are lovely downland grasses, particularly the attractive *quaking grass* with its purplish triangular-shaped spikelets shaking in the wind on slender stalks. This wonderful array of wild flowers attracts many species of butterfly including the *silver-spotted skipper*, the *Duke of Burgundy fritillary*, together with the *brown argus* which is occasionally seen on nearby Oxenbourne Down.

The *brown argus*, with a wingspan of some 28 millimetres, is dark brown with bright orange half-moons around the edges of its wings. The female is slightly larger and lighter coloured than the male. She lays her eggs singly on the underside of the leaves of the *common rock-rose, dove's-foot crane's-bill* and *common stork's-bill* plants, the latter being more characteristic of the Wealden greensand country, than of the Downs. The brown argus is a gregarious butterfly which is usually seen with a group of others, flitting to and fro among *bramble* flowers, *bird's-foot trefoil* and other wild flowers on which it feeds. The first brood of these butterflies can be seen in early May all along the South Downs but it is easier to spot them in July and August as the second brood, then, usually has higher numbers. Intriguingly, like other blue butterflies, (despite its name, this species is one of them), the caterpillars are often found with ants which milk them of secretions. The ants may in return be beneficial to the caterpillars in deterring predators.

The summit of Butser Hill is sadly marred by a telecommunications mast but the views from the top are magnificent. To the south the sea sparkles off the inlets around Portsmouth, and more distant the Isle of Wight lies resplendent; to the north-west, are green fields and farming country, and close by, the church of East Meon, standing slightly apart from the village, looks more like a cathedral with its massive central tower. From the heights of this hill no fewer than eight spurs with steep-sided combes taper down from the plateau. Their flanks are often mantled in trees, dark *yew, ash* and *hawthorn*, with here and there silver flashes of *whitebeam*. Sheep, carefully grazing the well managed grasslands, show up as little flecks of white on the wide stretches of green turf. For centuries Butser Hill has been their home under the control of the shepherd and his dog, and the sound of their bells blended with the trilling songs of *skylarks* high in the sky. Now the bells sound no longer, unneeded by the shepherd on his motorised transport, and the noise of thundering traffic down the A3 is always present as an unfortunate distraction. However, the *buzzard* still soars aloft on wings that hardly stir, and as I write a newcomer has become more evident, the *red kite*, but more of this bird of prey a little later. The sight of a pair of buzzards over the graceful flanks of Butser reminds me of the lines of Martin Armstrong in his poem 'The Buzzards':

> Serenely far there swam in the sunny height.
> A buzzard and his mate who took their pleasure
> Swirling and poising idly in golden light.
> On great pied motionless moth-wings borne along,
> So effortless and so strong,
> Cutting each other's paths, together they glided,
> Then wheeled asunder till they soared divided
> Two valleys' width (as though it were delight
> To part like this, being sure they could unite
> So swiftly in their empty, free dominion),
> Curved headlong downward, towered up the sunny steep,
> Then, with a sudden lift of one great pinion,
> Swung proudly to a curve and from its height
> Took half a mile on sunlight in one long sweep.

As on so many hilltops of the South Downs, history is all around and Butser is no exception. Here are ancient trackways, field boundaries, lynchets, earthworks and burial mounds. These scorings of ancient man leave you in a deep meditative mood, as you descend by way of Limekiln Lane, skirting the great combe of Grandfather's Bottom to the west of the summit; what further thoughts these names conjure up in a mind already filled with the intensity of this great hill. Dusk is fast approaching and the rays of the sun show briefly through darkening clouds, lending a comforting softness to the hillside. Looking back at the great mass of Butser Hill the lengthening shadows show up the lines of raised earth that are so common on downland slopes. These small earth banks running roughly parallel along the contours of the slopes are created by a phenomenon known as *soil creep*. Over hundreds of years the action of rainfall on the hills washes down the soil, which then forms in ridges somewhat like the ridges on a sandy shore shaped by the action of sea and waves. On the Downs these ridges are often made larger by the animals, especially sheep, who tend to compact them into paths as an easy way to traverse the slopes.

Hen Wood, just to the south and east of the lovely Meon Valley, is our next destination. The South Downs Way passes nearby and within the vast environments of this remarkable mixed wood are many tracks and open spaces. As you wander its leafy expanse, a *jay* close by startles you with its penetrating raucous call; 'straak'.

Jays are very common in deciduous woodlands where acorns and beechmast are an important food, and they are also found on farmland where copses and tall hedgerows are their preferred habitat. In recent years this bird has moved into suburban parks and other more urban areas where there are plenty of mature trees. About 32 centimetres long, the jay is one of our most handsome birds, easily recognised by its pinkish body, white rump and blue patch on its wings. In spring the sound of its harsh call sends a chill through smaller woodland birds, for like the magpie, it does a great deal of harm to them by eating their eggs and young fledglings. However, in the autumn, it hoards up acorns and other nuts in hideaways like squirrels, but sometimes these are forgotten and this helps to spread the trees when they germinate. For the greater part of the year the jay is a noisy sociable bird but as nesting time approaches, it becomes shy and furtive, each pair gliding quietly through the trees to build an untidy nest of sticks and roots, typically near the top of a *conifer, oak* or *sweet chestnut*. The five to seven green eggs, spotted and freckled with light brown are laid from April to June. Earlier in the last century, jays were shot as they were regarded as pests by the gamekeepers, but with the decline of gamekeeping generally, this species increased and has now stabilised with a population for Britain and Ireland of some 160,000 pairs with 9,500 pairs in East and West Sussex.

Another bird you may catch a sight of, if you are lucky, is the tiny fragile looking *treecreeper*. It is easy to overlook, as its sombre, buff-streaked plumage matches the bark of the trees where it spends its life climbing in search of insects and their larvae. Watch it as it climbs up a tree trunk, spirally in short bursts, with its stiff tail pressed closely up against the bark, and its long, thin curved bill picking out prey from cracks and crevices. The treecreeper's weak, high-pitched voice, 'tsee-tsee-tsee-tsizzi-tsee', can be heard in the quiet of the woodland throughout the year, but more often in the early spring. It makes its nest in a hole or crevice in a tree or behind a loose piece of bark still clinging to a decayed trunk. The nest is made of twigs and dead grass, lined with wool, moss and feathers; the six to nine white eggs, speckled with dark brown, are laid from April to as late as July. In the autumn and winter, treecreepers often join flocks of tits as they search for food amongst the trees. There are few more charming sights than that of *long-tailed tits* and *blue tits* flitting through the last brilliantly coloured leaves of the woods on a fine, clear autumn day, while a solitary treecreeper runs up and down and around the tree trunk and its branches.

Hen Wood has a wonderful ground flora but one plant in particular is distinctive; the *narrow-leaved helleborine*. This is a similar plant to the *white helleborine* described earlier, but its flowers are a much purer white and the flower spike is quite distinct from the rest of the plant. The flowers are also closer together, are more open, and the contrast between them and the fresh green, long, narrow leaves, gives this orchid a charming and elegant appearance. It is a rare plant found on or near the South Downs in Hampshire with occasional appearances in West Sussex. Here in the south, it favours *beech* woods on chalk, although in the north of England, in Wales and in Scotland, there are scattered patches in mixed woodlands of *oak* and *ash* on limestone. This species flowers in May slightly earlier than the white helleborine and seems to prefer the more open areas of the woodland margins. Other plants associated with it are *dog's mercury* and the graceful *sanicle* with its head of minute white flowers, smooth unbranched stems, and five-lobed leaves with toothed edges. The *white helleborine* also grows in the area, and occasionally hybrids between it and the narrow-leaved species have been found.

One of the great botanical discoveries of recent times in this general area is that of the very rare *red helleborine*. Some 15 years or so ago a party of birdwatchers came across this beautiful orchid growing in mixed woodland on the chalk; previous to this discovery it was known only in Gloucestershire and a single site in the Chiltern Hills. There was an old record for the red helleborine near Arundel, and who knows, with the felling and regeneration of our woodlands, there is just the possibility that it will be found elsewhere. It is very distinctive with up to fifteen (usually about five) beautiful rose-pink flowers appearing in June on a single stem from 20–60 centimetres high. The stem carries a few scattered, dark green narrow leaves along its length. The red helleborine often fails to flower and only the stem and leaves appear, possibly because of a lack of light when the wood becomes too overgrown. On the site in Hampshire, some trees have been cleared to allow more light onto the plant and there are some signs that this is having a positive effect.

Old Winchester Hill, with its ancient fortress and nature reserve, dominates the area just south-west of Hen Wood, and this is our next port of call. In this most westerly part of the South Downs,

Old Winchester Hill, to my mind, comes second only to Butser Hill for beauty and symmetry. This hill was part of the large dome of chalk that once spanned the Weald, and its summit is now enclosed by another of the great Iron Age hill forts, built over 2,000 years ago, with commanding views all around. The ramparts surrounding the 14-acre oval-shaped fortress are up to 5 metres high in places and at each end are well defined entrances. At the eastern end, two depressions are evident which may have been guardhouses, and numerous burial mounds are scattered within and outside the fort. Stand on top of the rampart and its surrounding defensive ditch one day, and soak in the primitive atmosphere of the place. Little disturbs your peace, just the rustle of many downland grasses at your feet and the flowing songs of *skylarks* high in the air above. To the south, across rolling countryside, now heavily cultivated with rows of wheat, is the coast and the great metropolis of Southampton; to the west is the lovely valley of the river Meon and on the skyline, the nature reserve of Beacon Hill. Just below you to the north is a lovely combe valley, lined with dark *yews* which blend in so beautifully with the light green of *beeches* around them. Steep slopes of chalk grassland and areas of scrub combine to make this a superb place for wildlife. Not surprisingly in 1954 the area was bought by the then, Nature Conservancy Council and is now a national nature reserve managed by English Nature.

There are some wonderful birds to be seen around Old Winchester Hill, many just passing through, but a significant number choosing to breed in the rich variety of habitat. There are also over 200 flowering plants and some 35 species of butterfly.

As you walk from the car park along the ridge to the hill fort in late April, you may well hear from a nearby thicket, the enchanting song of the *nightingale* with its deep bubbling and liquid notes. This somewhat skulking and shy bird, with almost uniform brown plumage, is more often heard than seen, and arrives here in the spring after a 6000 kilometre migration from Central Africa. It is not known to nest on the reserve, so the ones you hear may be just passing through, possibly to breed in woods bordering the Meon Valley further west. The nightingale prefers to take up residence in thick scrub, preferably in damp, lowland areas, but it is also found nesting in scattered localities in downland scrub. It builds its nest of dead leaves, lined with hair and grass, on or just above the ground, well hidden in nettles or low bushes. Forty years ago the nightingale was commonly heard from lay-bys beside the A27, where woods were close alongside, particularly between Chichester and Arundel; but now they are heard no more. The huge increase in traffic noise and the pounding vibration of heavy lorries on that road, now widened into dual carriageway, has clearly disturbed the nightingale greatly, and it has retreated a kilometre or so into the woodland, where the noise of traffic is only a distant hum, deadened by the tree foliage. You will know, of course, that the nightingale is famed for its singing at night, but not everyone is aware that it sings during the day just as often. I am sure you will hear it at some time on your wanderings over the South Downs. John Keats described it so beautifully in his 'Ode to a Nightingale', as he tells us of the:

> ... light-winged Dryad of the trees,
> In some melodious plot
> Of beechen green, and shadows numberless,
> Singest of summer in full-throated ease.

A *buzzard* quarters the sky above you and if you are patient and just a little lucky, you may catch sight of a much rarer bird of prey, the *red kite*. The story of this bird of prey in Britain is an interesting one. In mediaeval times, red kites were numerous all over the country and they were well known by Londoners who saw them catching *rats* and *mice* and feeding off carrion on the refuse-strewn streets. In the eighteenth century red kite numbers started to decline, brought about by the gradual cleaning up of towns, and the introduction of refuse disposal and underground sewage systems. This decline was aided by the relentless persecution of birds of prey by gamekeepers and farmers who treated them as vermin. It is sad to think that in 1777 the red kite was still breeding in Grays Inn, London, but by 1870 it was extinct as a breeding species in England and by 1900, in Scotland as well. Undoubtedly the onslaughts of egg collectors and taxidermists aided the demise of this bird, but luckily a few pairs held out in the wilder mountains of central Wales where between 1905 and 1945 some five or six pairs built nests. Since the Second World War their number gradually increased, although for a short while in 1954–55, myxomatosis in the *rabbit* population reduced the numbers of breeding pairs from ten to six. In the early 1990s it was decided to introduce kites of Spanish and Scandinavian origin to selected parts of England and Scotland, and this has been a great success, particularly in the Chiltern Hills near Oxford, where one can see them flying low over the fields next to the M40 motorway.

The red kite is about the same size as a buzzard, but is lighter in colour, with large whitish patches on the underside of its narrow, strongly-angled wings. It has reddish-brown upper parts, a streaked breast and a long, deeply forked chestnut tail. It makes its nest of sticks, lined with moss, wool and grass, liberally decorated with bits of paper, cloth, plastic bags and other oddments; a characteristic of this species. The nest is more often constructed in the branched fork of a tall tree, usually on the foundation of an old nest of a *crow* or *buzzard* and the same nest site may be used year after year. The two to four eggs are laid in early April and the young hatch out after about 31 days. The young grow rapidly, and leave the nest after a period of about 65 days and are dependent on their parents for a further two or three weeks. Breeding kites hunt for food from dawn to dusk with a peak period in the late morning. They soar and weave all over open country in slow effortless flight on bowed wings, their tails twisting to and fro, acting as rudders. When carrion is sighted the kite gently descends in circles and alights some distance away before walking or flying to the carcass. If a live prey such as a small *rabbit* or *vole* is spotted, then the kite commences a steep dive with closed wings directly onto the unwary victim. Kites also feed on small birds and insects and they like to be close to rivers which provide fish, and water meadows which provide frogs and toads. They have a close association with man, where they visit rubbish tips and have learnt that roads provide carrion.

The red kite is a passage migrant from mainland Europe and is sometimes seen in the spring and autumn. However, over the Meon Valley and near Butser Hill, red kites have been seen in increasing frequency and as I write a pair are overwintering in the area. One has a wing tag identifying it as originating from those breeding in the Chilterns, and I would not be surprised if by the time this book is published, there will be confirmation that this bird of prey is indeed breeding here. I hope you will see this noble bird, unmistakable in its buoyant, yet effortless flight, soaring above the woods and fields around Old Winchester Hill. Conceivably, their numbers will increase as they breed successfully; what a bonus that would be for our dedicated and hard working conservationists.

Many other birds may be seen on this magnificent reserve; those actually breeding here include the *chiffchaff, whitethroat, nuthatch, sparrowhawk* and *yellowhammer*, while others like the *golden oriole, hen harrier, merlin* and *osprey* are occasional winter visitors, or are just passing through on migration.

Of the butterflies on the reserve, perhaps the most striking and beautiful is the *purple emperor*. This magnificent butterfly, seen here only recently, is the largest woodland species in Britain with a wingspan of over 7 centimetres. Only the male has the distinctive, purple iridescent sheen on its wings, while the female is much browner in colour. Both have conspicuous white patches on their wings, together with eye-spots, encircled with brown. Purple emperors spend most of their time around the top of the tree canopies, particularly of *oak*, and only come down to feed on the nectar of flowers in the morning; by noon they will be back up into the tops of trees where they establish their territories. In late summer the female lays her eggs singly on the leaves of the very common and widely distributed *goat willow*, and these hatch out after about two weeks, the caterpillars feeding on the willow leaves before hibernating. In the following spring they start to feed again, before turning into chrysalids in June or July; after about 18 days the adult butterflies emerge in their full glory. There is only one generation and the best time to see them is in July and August along the paths of deciduous woodlands in the morning. Purple emperors are more common in the Queen Elizabeth Country Park, but you are more likely to come across them at Old Winchester Hill because it is so much smaller in area than the park.

The purple emperor is attracted to carrion on the ground where it can be seen feeding for long periods. In the nineteenth century it was a common practice for naturalists to place rotting bodies of animals on the ground to coax these spectacular butterflies down from the treetops; then using nets on the end of poles, up to 9 metres long, to catch them. The purple emperor is an elusive butterfly, and despite its size and colouring, is probably under-recorded, and because of this it may be more common than at first thought. Keep your eyes open in those crucial summer months as you could see them just about anywhere in the deciduous woods of the western Downs.

Another handsome butterfly to be found in open places on this reserve is the *dark green fritillary*. This large butterfly, with a 57-millimetre wingspan, is not unlike the *silver-washed fritillary* described earlier, but it can be distinguished by the brighter silver and green markings on the undersides of its hind-wings. This species has a very swift and powerful flight as it soars and skims over downland flowers, especially *thistles* and *knapweed* which it loves. It is best seen in July and August when it settles for a moment on these flowers to take nectar. The female lays her eggs singly on the leaves and stems of *violet* plants in August, and after hatching, the colourful blue, red and green caterpillar devours its eggshell and then immediately hibernates until the following spring. It then emerges to start eating the leaves of the host plant, eventually turning into a chrysalis in May. The adult butterflies appear in June and live for 20 to 35 days.

Other butterflies that you may see on Old Winchester Hill include the *brown argus, white admiral, green hairstreak, large skipper*, and *silver-spotted skipper*.

One of the typical sounds of summer is the chirping of grasshoppers. Here, in late June and July, amongst the *timothy grass* and other wind-blown grasses on the summit, you may hear the unique undulating song of the *striped-winged grasshopper*. This 2-centimetre long grasshopper gets its name from the white stripe on the forewing; its presence on Old Winchester Hill is indicative of an old and flora-rich, chalk downland.

Many of the orchids are represented here, including the *fragrant, fly, common spotted* and *pyramidal*; this last named with its pyramid-shaped bright pink flowers being particularly conspicuous on the steep slopes above the combe in July and August. *Cowslips* cover the area in the spring, and many other yellow flowers appear in May and June such as the *common rock-rose, lady's bedstraw, kidney* and *horseshoe vetch*, the last named being an important food source for the *chalk-hill blue butterfly* which is very common in July and August around the area of the hill fort.

One interesting plant found on Old Winchester Hill and in many places on the South Downs, particularly where the grass is short, is the *eyebright*. This herb has bright white flowers with beautiful markings of purple and yellow which appear from July to September. In short grass, the species may obtain a height of only 5 centimetres, but where the grass is long it may reach to 20 centimetres or more. It has deeply toothed, dark green oval leaves which are hairy underneath. Eyebright is a partial parasite feeding on the roots of various grassland herbs and robbing them of sustenance. There are many different varieties of eyebright, and their Latin name is derived from the Greek word, *euphrasio*, meaning to delight or gladden, an allusion to the pleasing colours of their flowers, or to the joy of having eye complaints cured with their extracts. In mediaeval times eyebright potions supposedly cured dimness of vision and short sight. Herbalists in the seventeenth century prepared a powder from the plant which was used for brightening the eyes.

The *round-headed rampion*, so common on the Sussex Downs, also grows here on Old Winchester Hill, where their striking dark blue flowers make an attractive sight as they sway back and forth in the wind. Near the main car park to Old Winchester Hill in May and June you may see the tall and bristly plant of *common comfrey* with its long oval leaves and drooping, funnel shaped, usually pale yellow flowers; sometimes their colour being red or purple. This plant grows particularly well by streams and roadside ditches and was once a common cottage garden plant, for it was used not only in the kitchen but also as a healing poultice for sprains and abrasions. Its country name was 'knitbone' and the roots were lifted in spring, grated and made into a mash that set solid; an early forerunner of the plasters and bandages used today. In the Middle Ages, comfrey was highly regarded by monks who grew it and used it to treat sick and injured travellers. In the kitchen, young comfrey leaves were used and eaten like spinach, served in a cheese sauce or made into soup. The roots, once dried, were roasted and ground up and used as the basis of a hot drink, like coffee. In addition, infusions made from the plant were used until very recently as herbal remedies against stomach upsets, lung troubles and coughs, but this practice is now discouraged because it has been found that the alkaloids in them can cause serious liver damage.

It is time to leave this nature reserve at Old Winchester Hill with its history and remoteness and descend into the broad alluvial flood plain of the river Meon, where you will find many undisturbed and tranquil places. As you follow the track that leads down from the top of the hill and across a disused railway line, a *stoat*, making short work of a *rabbit*, is startled by your sudden presence and bounds away into nearby scrub. The river Meon, its clear sparkling waters once famed for their use in wool washing and dyeing, and now for its trout fishing and watercress, originates from springs just to the south of East Meon. As it makes its way to the sea just west of Gosport, it passes through diverse habitats of rounded chalk hills, wooded valleys, country estates and cultivated fields. Sometimes wide and slow-flowing, sometimes narrow and racing over stones and smooth pebbles, this delightful river is always alluring to the walker; and to the fisherman seeking not only his prey, but more often peace and quiet, it is held in high esteem. Although Alfred Lord Tennyson wrote his poem 'The Brook' describing a stream in the Lincolnshire Wolds, it could well apply to this delightful river in East Hampshire:

> I come from haunts of coot and hern,
> I make a sudden sally
> And sparkle out among the fern,
> To bicker down a valley.

By thirty hills I hurry down,
Or slip between the ridges,
By twenty thorps, a little town,
And half a hundred bridges.

Till last by Phillip's farm I flow
To join the brimming river,
For men may come and men may go,
But I go on for ever.

I chatter over stony ways,
In little sharps and trebles,
I bubble into eddying bays,
I babble on the pebbles.

With many a curve my banks I fret
By many a field and fallow,
And many a fairy foreland set
With willow-weed and mallow.

I chatter, chatter, as I flow
To join the brimming river,
For men may come and men may go
But I go on for ever.

The river is the haunt of the *kingfisher*, that brilliant bird of azure blue-green and orange, with white throat and sharp-pointed bill. If you walk quietly along the banks you may see one sitting solemnly on an alder branch as it peers down looking for small fish. Suddenly, with a clear 'kee-kee', it flies upriver just off the water; fleetingly just a flash of iridescent blue. The kingfisher makes its nest in the river bank where it bores a hole about a metre deep, at the end of which it lays five to eight pure white eggs on bare earth in March or April. Young kingfishers are fed on a diet of minnows and other small fish, as well as insects such as dragonflies and water-beetles. Often, the nest itself becomes badly fouled with the bones and other remains of this food.

Unfortunately water pollution has depleted the numbers of this beautiful bird and it is now seldom seen along the main rivers of southern England; however it can still be spotted on quieter stretches of the smaller and largely unpolluted streams and rivers such as the Meon. This bird is also badly affected by cold weather and their numbers were very much reduced after the winters of

1962–3, 1984–5, and 1985–86. Many of the birds move to the coast in winter where the warmer sea temperatures keep the nearby beaches and estuaries from freezing and enable them to survive. Seemingly, the spate of warmer winters of the 1990s has enabled this bird to increase in numbers; there are some indications that this has happened.

The warmer climate of recent years has brought to the shores of southern England increasing numbers of that elegant and beautiful bird, the *little egret*. Before 1958, it was a rare summer visitor from the Mediterranean, being recorded in Britain less than fifty times; now it is a regular visitor and is seen throughout the year in river valleys and coastal estuaries. The first record of it breeding was in Dorset in 1996, where one pair produced three young, and there is little doubt that it is now breeding in other localities. The little egret, about 56 centimetres long, is a small snow-white heron, with long slender black bill, black legs and yellow feet. Towards the end of the eighteenth century and the early part of the nineteenth century the silky plumes of this bird and other members of the egret family were highly sought after for the adornment of women's hats, particularly for weddings. Demand outstripped supply and it was not long before the little egret population was almost wiped out, particularly in the deltas of the Nile and Volga. In 1889, the RSPB was formed to campaign against the feather trade. This was successful, and Queen Victoria ordered the banning of egret plumes in military headdresses; by 1921, further legislation banned the import of plumes into Britain. The feather trade, for which London was then the global centre, declined as plumed hats became unfashionable. As a result, little egret numbers began a slow recovery.

Little egrets usually nest in colonies among trees and bushes in woods, but sometimes the nests are in marshes and in open country. You will see single birds or groups of them just about anywhere close to water in the Meon Valley, but particularly along the river's edge and beside pools and lakes. Watch for their slow graceful flight and in the breeding season listen out for their croaking calls, 'kark-kark', and bubbling 'wulla-wulla-wulla'.

Another bird you may come across in the valley, particularly on reedy pools and small lakes, is the *little grebe*, or *dabchick*, as it is commonly known. About 25 centimetres long, this somewhat shy, retiring bird, is the smallest of the British grebes and can be recognised, not only from its size, but also from its overall dark brown colour, with chestnut cheeks and throat; in the winter it is much paler and without the chestnut colouring. It makes its nest, a flimsy raft of water weeds, amongst reeds, or in some sheltered spot in the open. Both parents take turns in incubating the eggs and like the *great crested grebe*, they cover the eggs with vegetation before leaving them. The young chicks flourish in the protection and cover of waterside vegetation and they grow quickly, feeding off a diet of insects, larvae, tadpoles and other pond life. Their favourite perch is on their parent's back and what a delightful sight they make then, tucked under their mother's wing. In the winter, whole families of little grebes make their way to larger lakes and reservoirs, seldom remaining in their breeding quarters. You will need patience and not a little stealth to observe these birds, as at the first sign of danger they submerge and head for the nearest patch of aquatic vegetation, to hide.

The waters of the river Meon are possibly not extensive or substantial enough to be favoured by the *otter*, which is now spreading across the river valleys of Southern England, having been introduced in recent years from captive breeding stock. However, since several pairs are now known to thrive on the river Itchen, not far away between Cheriton and Winchester, it is only a matter of time perhaps, before their offspring make their way towards the Meon Valley. The otter, a large carnivore belonging to the weasel family, was a widespread and fairly common mammal of rivers and streams throughout the country before the Second World War. I remember seeing it often in the Cuckmere Valley around Arlington in the early 1950s; but because their favourite food is fish, hunting and trapping by water bailiffs along salmon and trout rivers brought a sad decline in their numbers, until by the 1960s it was virtually extinct in Sussex, Hampshire and many other southern counties. Pollution of our rivers by new pesticides in the late 1950s was also a factor in the otter's decline, because these chemicals accumulated in the bodies of fish and eels and affected the fertility of whatever ate them. Its survival is not helped by the rapid spread of the *American mink* into the wild from farms, where it is bred for its fur. The mink soon establishes itself in the wild and being a voracious and aggressive animal it easily drives away the otter from its natural habitat.

The otter grows to about 80 centimetres long with a thick tapering tail of about 40 centimetres which it uses as a rudder. Its chocolate-brown coat consists of fur, under which is a thick layer of short hairs, some 20,000 of them per square centimetre; not surprisingly this makes the coat completely waterproof. With a beautifully streamlined body and webbed feet, the otter is a formidable hunter diving for fish and eels at night when they are easiest to catch. It requires quiet rivers with many tree roots protruding from their banks into which it makes a tunnel for its breeding nest, called a holt. Usually there are only one or two cubs in the year, with perhaps three or four if food is plentiful; these cubs may be produced in any month of the year, there being no definite breeding season. The young cubs venture out of the holt after about two months and remain with their mother for about a year. As soon as they leave the holt and can fend for themselves, albeit with some help from their mother, the male or dog otter leaves the family to live on its own. Otters have a lifespan of about four years and only about two litters are produced in this time, and it is not surprising, therefore, that they find it difficult to increase their population. Every year many of them are killed, particularly at night, on our increasingly busy roads, and this remains one of the biggest causes of their premature death; in some places local authorities have built special tunnels under roads near otter populations and this has proved effective. Otters are shy and not often seen, the best indication that they are around is to find their droppings, called spraints, and their five-toed webbed tracks in mud or sand. They mark their territory by depositing spraints at selected places that can be found by other otters and act as a 'keep-out' warning sign. These places are usually sheltered, such as on a rock under a bridge or overhanging bank, so that they are not destroyed by the sun and rain. Let us hope that the efforts of conservationists to increase their numbers and make them more widespread will be successful, and that once again they will be seen on our southern rivers and particularly here in the Meon Valley.

It is time to leave this valley as the autumn tints begin to show along the way, with *beech* and *birch* trees are faintly tinged with yellow and bramble leaves touched with red. *Fungi* begin to show in the woods and on the hillsides, and *martins* and *swallows* begin to gather in groups as they prepare for their long migration southwards. You are excited to see a *red kite* slowly quartering the distant woods near Old Winchester Hill, apparently intent to overwinter in the area. You know that a harsh winter will be a deterrent to it in the future and may also drive the *little egrets* further south to the coast or even back to the warmer parts of the continent. As winter passes and the days lengthen with the approaching spring, so *frogs* and *toads* start emerging from hibernation and begin to breed in damp places. Only their phenomenal breeding rate ensures their survival as their eggs, tadpoles and young are preyed upon by fish, newts, water birds, snakes, rats and hedgehogs, forming an important part of the food chain. Sombre *alder* trees with their dull purple catkins give colour to the landscape and down the lane a brown and grey *dunnock* bursts into song 'wee sissy-weeso, wee sissy-weeso'; remarkably loud for such a small bird.

With great reluctance you say farewell to this wandering river and take the South Downs Way that passes through Exton to the nature reserve of nearby Beacon Hill, owned and managed by English Nature. Here a mixed woodland which includes *birch*, *ash* and *beech* is flanked by steep slopes of chalk grassland giving a variety of habitats. Typical downland flowers like *common rock-rose*, *wild thyme*, *marjoram* and *scabious* are to be found, together with associated butterflies such as the *brown argus*, *common blue*, *meadow brown* and various *skippers*. A flash of colour catches your eye as you walk on the edge of the wood and quickly you identify a pair of *goldfinches*, as they settle for a moment on a thistle head to eat the seeds. The charm of goldfinches is in their beautiful colouring; bold markings of brown, black and yellow with brilliant scarlet faces and black and white heads. Their flight pattern is quite distinctive, undulating and almost butterfly-like as they flit from thistle to thistle, twittering a liquid 'swett-witt-witt'. They are gregarious birds and often gather in large numbers, especially outside the breeding season. The goldfinch builds a neat little, cup-shaped nest of moss, wool, grass and lichens, lined with feathers and hair. A typical site for the nest is near the end of a *horse chestnut* or *sycamore* bough or in the fork of a fruit tree in an orchard. Some 80 per cent of our goldfinches migrate to mainland Europe in the winter and those that remain form small flocks, often mixed with other finches, to feed predominately on the seeds of alder, birch, thistles and teasel. John Keats in his poem, 'I stood Tip-toe upon a little Hill', so aptly describes the goldfinch:

> Sometimes goldfinches one by one will drop
> From low-hung branches; little space they stop;
> But sip, and twitter, and their feathers sleek;
> Then off at once, as in a wanton freak;
> Or perhaps, to show their black, and golden wings,
> Pausing upon their yellow flutterings.

The South Downs Way here zigzags its way across country by minor road and track, sometimes north-west and sometimes south, but always progressing westwards towards Winchester. Some 6 kilometres beyond Beacon Hill you come to an attractive part of the Hampshire hills, Gander Down. This is an area of mainly agricultural downland with fields that are bounded by areas of rich grassland. Trackways give public access to the Down and in places they pass through shady copses and tall trees; as you walk these paths you will see much to interest you. Take the track named Honey Lane which branches off from the South Downs Way some 400 metres south-east of Ganderdown Farm, and then takes you north-east on a circular walk to rejoin the South Downs Way about 700 metres north-west of the farm. This lane is bordered by tall *oak* trees whose age clearly indicates that this was once used by horse traffic for generations, well before the arrival of the tractor. In spring and summer the grassy borders of this lane are richly coloured by many wild flowers and butterflies. You are almost certain to see a *comma* butterfly along this walk, particularly in April and May or in August or September. This orange-brown butterfly with black markings gets its name from the distinct white comma mark on the underside of the wing. It is the only British butterfly that has a jagged edge to its wings and when it rests with its wings folded. These edges and its mottled coloration, make it look just like a dead leaf which allows it to hibernate undetected over winter. However, its marvellous use of camouflage does not end there, for the caterpillars which feed on hop, nettle or elm leaves look just like bird droppings, giving them some safety from predatory birds. The comma tends to be solitary and delights in basking in the sun with wings spread, sometimes for hours. It favours the nectar of *bramble* flowers, *thistles*, *knapweed* and *hemp-agrimony*. The comma population fluctuates; about 35 years ago they were quite rare, and now they are a common sight, not only in the countryside, but also in urban gardens, particularly in the autumn when they like to feed on rotten fruit.

Honey Lane takes you downhill for just over a kilometre and a half, and just beside a large hay barn, you turn left and westwards, uphill along a metalled lane. Some 180 metres further on you leave the lane through a narrow gateway that leads into shady woodland. Here, low beech boughs form a natural archway over the shady woodland track, and in spring this place under the great trees is carpeted with *bluebells*. As the path emerges from the wood, there is a large *rabbit* warren among tree roots in the bank. Rabbits abound in the fields all around and it will be interesting to see if they become afflicted with a new disease that now affects other rabbits, as at nearby Old Winchester Hill and indeed many places all over Britain. The disease, which was first noticed in Devon in 1994, is called *haemorrhagic* and destroys the liver and lungs of rabbits leaving them motionless and unable to walk. As a

217

result they make every effort to return to their burrows and then die. The spread of haemorrhagic is patchy as it seems to affect some rabbit colonies but not others, even sometimes when the colonies are close to each other. Only time will tell how long it will last and what effect it will have on the overall rabbit population and on their predators.

The views from the path are extensive, covering broad fields of cultivation interspersed with dark green woodlands. Across the valley the outline of an Iron Age burial ground is just visible and reminds you of how Hampshire's gentle climate and rich soil attracted Neolithic farmers to the area more than 4000 years ago. As the population increased, so hill tops, such as nearby Old Winchester Hill and St. Catherine's Hill, were fortified to defend resources against unfriendly neighbours and from invaders further afield.

The path joins up again with the broad track of the South Downs Way. In high summer when the flowers are at their best you will see the orange-coloured *small skipper* butterfly darting from one flower to another with fast, whirring flight. This common species is found in almost every patch of rough pasture particularly where the grass is tall, and you will certainly have seen it on other parts of the South Downs. At rest, the skipper usually keeps its forewings half closed and its hind-wings out flat; this is known as the 'skipper position' and is common to that family of butterflies. You may also come across a very similar species here, the *large skipper*. This differs from the small skipper in having a more solid-looking body and by having a slightly chequered appearance on both sides of its wings. If you can get close to it you may notice that its antennae are distinctly hooked at the tips whereas those of the small skipper, in comparison, are just gently curved. The males of both species have dark scent marks showing as lines on the upper fore-wings, and these have an important function in attracting females. The caterpillars of both species feed on grasses and spend the winter inside cocoons, spun within the blades of grass, before emerging in the spring to recommence feeding. After pupating the adult butterflies emerge in June and are on the wing until the end of August.

The South Downs Way passes by a small wood at Cheesefoot Head just to the east of Winchester. Pause here awhile, and gaze down at the deep and impressive valley called the Devil's Punch-bowl. The steepness of the downland slope here has saved it from the plough, and in the summer you will find a host of wildflowers such as the *clustered bellflower*, *fragrant orchid* and *frog orchid*.

Not far away you may come across yet another variety of the *bee orchid* designated *belgarum* after the name the Romans gave to Winchester. The labellum of this bee orchid is rounded with odd patches of brown and yellow and quite unlike that of the normal species.

As you cross the busy A272 road, the immensity of the broad sweeping hills and the open chalk lands of Longwood Warren to the south, become apparent. On the skyline, tree belts show up as dark smudges across a landscape made barren by featureless fields of agriculture. Once this land would have been open grazing and the haunt of the *great bustard* and *stone curlew*; sadly these birds have long since gone. In the eighteenth and nineteenth centuries most of such wide open spaces were hedged or fenced off as a result of the Enclosure Acts. These Acts were decreed as a result of improved agricultural and animal breeding techniques which necessitated crops and cattle being protected from wild herbivores and predatory carnivores. Landowners, too, sought this legislation so that they could fence in commons and open fields which was of course to be to the detriment of poorer people. It was not until the last half of the twentieth century that farmers, seeking efficiency from economies of scale in a highly mechanised agriculture industry, brought about the removal of many hedges and fencing, so creating a landscape of large open fields.

Our journey is nearly at an end and St Catherine's Hill lies just a few kilometres to the west, rising steeply from the Itchen Valley that lies to the south of Winchester. It is best approached by the track that leads from a car park just to the north of the hill itself. St Catherine's Hill was a centre of human settlement around 3,000 years ago, well before the founding of Winchester itself, and a fort was constructed on top of the hill in the third century BC by Iron Age Celts who saw the military significance of such a place, with its steep-sided slopes, and clear views all around. Within the ramparts and at the highest point, a chapel was built in the twelfth century and dedicated to St Catherine. This chapel was destroyed in 1537 following the Dissolution of the Monasteries in Henry VIII's reign, and in 1762 a clump of *beech* and *sycamore* trees were planted as a commemoration by the Gloucester Militia who were stationed nearby. Apart from its historic importance, St Catherine's Hill is a wonderful place for the naturalist and is managed on behalf of its owners, Winchester College, by the Hampshire Wildlife Trust. The route to the top is well signposted and as you pass through the gate at the bottom you will see two large *walnut* trees, probably descendants of those brought by the Romans some two thousand years ago. It is June, and before you start the steep climb up, you marvel at the beauty of thousands of *oxeye daisies* turning a nearby field almost white as their flowers sway to and fro in a light breeze. At the far side you notice several *roe deer* nonchalantly tiptoeing through the long sward stopping occasionally to feed.

The climb to the top is aided by steps cut into the chalky soil and supported with wooden planks. All around you are leafy boughs of *hawthorn*, *ash* and *beech*, which give you shade and help to deaden the noise of traffic thundering down the M3 motorway which cuts through the hillside just to the east. Beside the steps, the *dogwood* shrub puts up its slender blood-red twigs with short stalked leaves which grow opposite each other. These leaves are amongst the first to appear on the chalk downlands; pale green in early March, then turning a deep red shade in autumn to match the colour of the twigs. The small white four-petalled flowers appear in May and have an unpleasant odour. By October the flowers have developed into soft, round black berries which are swallowed by birds, which then distribute the seeds around the area so enabling the shrub to spread. The name dogwood has no association with dogs; more properly it should be called dagwood, because its branches were used by country folk to make what they called dags – skewers or pegs. In ancient times dogwood was used for arrowshafts.

The path rises steeply at the top and the remains of the huge rampart come into sight, a massive circular wall with a deep ditch that encircles the entire crest. The summit here is fairly flat and near its centre there is a huge maze dug into the turf, the origin of which remains a mystery but is not thought to have existed before the seventeenth century. Beyond the maze is a mound, situated amongst the clump of huge *beech* and *sycamore* trees, and is all that remains of St Catherine's Chapel. As you sit on top of this mound, look up at small patches of blue sky through the swaying branches. The wind always seems with you on the high tops of these South Downs and even here under tall trees it penetrates, and as it blows the brown leaves at your feet, a small *speckled yellow moth* is disturbed from its rest. This moth has bright yellow wings, spotted with dark brown and is quite common along woodland paths in Southern England in June and July. The eggs are laid on *wood sage*, *dead-nettles* and *hedge woundwort*. As you watch, this moth flies across the leaf litter to settle again some distance away. Your mind wanders back to the distant past; who were the people who worshipped here and how did they live their lives?

You emerge from the gloom of the woodland canopy and onto the undisturbed open grassland slope that overlooks Winchester itself. In summer this chalk downland has many of the plants and butterflies that we have seen all along the South Downs; flowers like the *harebell, dropwort* and *yellow rattle,* and butterflies such as the *brown argus, chalk-hill blue* and *marbled white,* to name but a few. Below you stretch the lush green water meadows of the river Itchen with their lines of willow, reed and dark rushes. In the Middle Ages these meadows were used in the winter and spring for the rearing of cattle and sheep, which returned in the summer to St Catherine's Hill and the Downs further east.

Beyond, standing out amongst the grey walls and red tiles of the city is the great mass of the cathedral itself, its tower looking silvery-grey in the sun. Watch for a while and see the subtle play of the shadows on its ancient stones, changing as the earth revolves on its axis and about the sun. Directly below you is the eighteenth-century Itchen Navigation Canal which linked Winchester with Southampton. Alongside there was once a road and railway, but these are now long gone and St Catherine's Hill is once again reunited with its river and water meadows. To the south the river Itchen shows as a silver streak, and to the west there are wooded ridges set upon the great chalk landscape that stretches away into Wiltshire.

This seems a fitting place to end one's journey across the South Downs for here there is a natural beauty of open grassland and tall trees, not perhaps as dramatic as the great chalk cliffs of Beachy Head, but still leaving you with a deep sense of peace and contentment. Here, too, the ancient city below, reminds you of the great history all around, and of a past that has always seemed so close throughout your travels over these beautiful hills.

The South Downs

> Chalky white cliffs,
> Stare, erect and proud,
> Watching waves hurl,
> Salt spray, thunderous and loud,
> Seagulls soar,
> Their haunting cry,
> Reminds us of fossils,
> Ancient sites – time gone by.
> Velvety green folds,
> Whose grass protects;
> Beautiful flowers,
> Animals, insects.
> A nature diverse,
> Historic small towns,
> The beauty so untouched,
> Is the charm of the South Downs.

> Jane Coulcher
> 1st January 2000

St Catherine's Hill, Winchester

EPILOGUE

Harting Downs

How does one conclude a book which attempts to cover so much of what there is to see on the South Downs? Our journey has taken us from the gleaming white cliffs and thundering sea of Beachy Head, to the history and relative calm and tranquillity of St Catherine's Hill. Between these two points, particularly to the east, are wide open spaces of chalk grassland, which for me, cast a spell of solitude, when past and present, light and land, all transcend and lift my mood of melancholy, creating a sense of joyousness and freedom. To the west are the wooded hills where at times, the deep, hushed, silent stillness of tall trees reaches into my soul and very being. River valleys, lush with vegetation, wind their way through the hills, and nestling, unsuspected among downland folds and trees, are attractive villages, descendants of ancient settlements whose kindly people still cultivate the fields and take care of the forests.

I am very conscious that I have barely touched the face of nature and have left out so much that is known to many; yet I hope my inadequacy in this respect is ameliorated by the descriptions of the wildlife that I have covered. Again it has not been possible to include an account of every nature reserve, every hill, wood or valley but I hope the places that I have mentioned appeal to the reader.

The South Downs in the most part are free for all to explore and enjoy, and for those who seek refreshment from the pace of life

223

and an escape from the stress of everyday toil, let them take the white chalk tracks up to the top of these hills. When winter mornings dawn crisp and clear, feel the frost rustle under your feet as it lies thick upon the hillside grasses, motionless like moonlight. As the sun rises, watch its brightness slide across the folds of the great escarpment. A *buzzard* soars upwards in ever-increasing circles as it catches an early thermal and then a *skylark* pours out its liquid song high up in the clear air. Abruptly the tranquillity is disturbed by the noise of heavy machinery starting up as workmen arrive to continue their widening of the road down below. Nearby a *woodcock*, disturbed from its early morning search for worms, flies to safety amongst newly-planted conifers. Suddenly, as you watch, hedgerows and roadside verges disappear before your eyes, you realise how vulnerable all this is to mankind's insatiable desire for progress.

Anyone who has seen, as I have, the changes that have taken place in our countryside over the past fifty years, cannot but conclude that if we go on as we are, then there will be precious little left for our successors to enjoy. Thank goodness this problem has been recognised and that many public and private bodies have been established to protect the countryside and its wildlife. However, we must always be vigilant to ensure that the relentless changes of our civilised society do not steal away our heritage in underhand ways and furtiveness.

By the time this book is published I hope there will be confirmation of *otters* breeding on some of the South Downs rivers. What a bonus it would be too, if *little egrets* and *red kites* were by then well established as nesting birds. Perhaps, despite my reservations, whoever reads this book in fifty or even a hundred years' time may be able to say 'the South Downs have not changed much since he wrote about them at the beginning of the last century except that some species have fortunately been re-established'.

Finally, I touch upon the spirit of the Downs and the place I love above all others, the Cuckmere Valley with its great 'Cathedral of the Downs', the parish church at Alfriston set upon flats of marsh and meadows of green grass; here one can sit beside the river at ebb tide and look into its still waters to stare mortality in the face. As the sun lowers through the mists of autumn and the lights of the church appear through stained glass windows, somehow you sense more strongly the spirit and life force that have been your companion amongst the beautiful hills and valleys of the South Downs.

POSTSCRIPT

WHERE HAVE ALL THE FLOWERS GONE?

Old Winchester Hill

Throughout this book I have touched upon some of the problems associated with the conservation of the South Downs as it relates to its natural history. In order not to detract the reader from the flow of the main part of the book, I thought it appropriate to deal with these problems; grazing, roadside cutting, rabbits and others, as a postscript. In doing so, I acknowledge fully at the outset that I am not an expert in conservation and I have no professional background in such work, but with over fifty years as an amateur naturalist I feel that my observations on what is amiss with our countryside, particularly the South Downs, are worth setting down. In doing so here, I realise that much of what I say is controversial, but I hope that it is constructive, using common sense and not the sometimes convoluted facts of so-called experts.

I begin with the problem of grazing which the reader who has patiently read the previous chapters will realise is of particular concern to me. For centuries much of the South Downs have been grazed by livestock, mainly cattle and sheep, and this prevented the spread of such shrubs as *hawthorn*, *bramble* and *dogwood*. Just after the last war the intensity of grazing decreased and large areas of scrub formed, where once there was a rich flora on open grassland. The establishment of conservation groups like the Sussex

Wildlife Trust and English Nature has done much to reduce scrub encroachment on sensitive areas with the aid, in many cases, of volunteers. That of course is a good thing, but what I am concerned about is what happens to the area after the grass and rich flora reappear. One hears the argument that the area must be grazed by sheep and cattle to prevent the scrub from reappearing. My contention is a simple one; if livestock are allowed to graze all the year round, albeit where there is sufficient grass available, then this all-important habitat will be devoid of its normally rich flora except for a few low-growing species like *eyebright*, *thyme* and *stemless thistle*. One need only look at the Downs just to the west of Eastbourne to see how increased grazing has virtually eliminated *cowslips* which only ten years ago covered the hillsides yellow in the spring. Again, on the Seven Sisters the once tall grass is now so short that wildflowers like the *round-headed rampion, field fleawort* and *carline thistle* together with many species of fungi, particularly *lepiota* and *horse mushroom*, are now seldom seen. Not only have the flowers gone, but birds also, like the *skylark* and *meadow pipit* which rely on long grass for cover and nesting. Walk on these hills at any time of the year and you will see hundreds of sheep and cattle, and of course this is true not only around Eastbourne, but also in many other 'environmentally sensitive' areas such as Mount Caburn where sadly the wind no longer rustles tall grasses on its downland slopes. The blame lies not with the farmer, for he must take advantage of the efficiency resulting from economies of scale, and graze as large a flock of sheep as the land can sustain.

I am well aware that some plants thrive in very short grass, as do some butterflies like the *silver-spotted skipper* and the *Adonis blue*. However, most flourish when the grasses have the benefit of spring growth. Surely it must be possible to fence off part of a rich habitat, at least from April to September, thus allowing plants to flower and set seed, and so allowing ground-nesting birds a chance to raise young, unhindered by wandering cattle and with some cover from marauding crows and other predators. Of course, such a system reduces the available grazing land for the farmer's livestock, and he would need to be compensated for the reduced numbers of cattle and sheep that he would be able to rear; but any conservation, if it is to be effective, has a cost, and this must be borne by those who gain, i.e. the public through the government of the day.

Many suggest that *rabbits* cause most of the destruction of chalk grassland. From my personal observation this is true to only a very limited extent in some small, well-confined areas. Some well meaning people go as far as to suggest that rabbits should be destroyed in

226

large numbers. I do not favour this solution as nature will in the long term resolve the problem in its own way; the latest disease of rabbits, described in the last chapter, may be such a process, and in any event the proliferation of rabbits does bring one bonus in the appearance of greater numbers of birds like the *buzzard* and perhaps the *red kite* that are attracted to an abundant food source.

Another problem that we face today is the cutting of the verges along our roadsides. The practice seems to be to cut all the vegetation right up to the roadside hedge or boundary fence even if it means cutting many metres of rich grassland habitat. By so doing, many of our wildflowers never have a chance to set seed since from my own observations the cutting is carried out regularly in the spring and summer growing months. I know of many roadsides which once held a rich variety of wild flowers but they have all disappeared because of roadside cutting. To emphasise this point, even as I write, an important colony of that rare plant, the *tuberous pea* has just been 'grubbed' out as roadside scrub is cleared (unnecessary in my opinion) on an area of downland just above Eastbourne on the road to Beachy Head. Hopefully this will reappear but it will then need positive protection from any further "tidying up" of the roadside. I have no argument with the necessity to show clearly the actual edge of the road or the requirement to cut back tall vegetation which obscures the driver's view, but surely it is not necessary always to carve such a broad swathe through an important wildlife site. Conservancy posts that have been introduced in recent years to mark areas of roadside where scarce plants grow, do have some effect in preventing their destruction by the roadside cutter, but many of these posts have disappeared and have not been replaced; in any case if only the first 60 centimetres of the verge were cut, then posts would not be necessary and there would be financial benefits to the local authorities. Our roadsides, too, would look much more attractive with wild flowers, rather than being littered with shreds of plastic, broken bottles and tin cans, as is so often the case.

The last point I wish to raise is the contentious issue of the National Park status for the South Downs. At present there are many groups involved with the protection and conservation of the Downs and their wildlife; the Sussex and Hampshire Wildlife Trusts, the Council for the Protection of Rural England (CPRE), English Nature, and the Sussex Downs Conservation Board, to name but a few. All do much to conserve, protect, and improve the natural beauty of these hills and to promote the public's enjoyment of them. Such bodies do much good work in meeting the broad aims and their own specific objectives, but many rely on financial support from public fund-raising and this, together with the lack of strong legal powers, inhibits them to some extent.

On the 29th September 1999, at the fiftieth anniversary of the creation of the first National Parks in England and Wales, the Deputy Prime Minister announced that he was taking the first steps in creating the South Downs a National Park. This announcement was welcomed by some, like the CPRE and the Council for National Parks, but others had reservations.

The main objections to National Park status seem to centre on the loss of local control of the countryside and the fact that many more visitors would be attracted to an already overcrowded area. Such objections are possibly founded on the problems associated with other National Parks, but many lessons have been learnt since our first National Parks were declared in the 1950s. New legislation in 1995 strengthened and improved arrangements for safeguarding and managing them and as a result, there is now a greater emphasis on conservation and the management is more democratic. In addition, the new Environment Act offers the opportunity for a new-style South Downs National Park Authority to be tailor-made to local needs.

National Park status provides the highest level of legal protection available in the United Kingdom and it would be a statutory duty of the Authority to conserve and enhance the natural beauty, wildlife and cultural heritage of the Downs. There would be the strongest opposition to major developments and the new Authority would be the overall strategic planning authority, although development control could be delegated back to local bodies.

Another obligation of the Authority would be to promote 'opportunities for public understanding and enjoyment of the special qualities of the area' and resources would be provided to manage the huge recreational pressures. The law specifically says that where there is conflict between conservation and recreation, conservation must come first. This is important and should do much to meet the objections of those who believe that a National Park means too many visitors. When the 1995 Act was passed the government made clear statements that National Parks should not be concerned with promoting numbers of visitors, but with the quality of experience they offer. Indeed, since the Norfolk Broads acquired National Park status, the Authority has transformed the quality and quantity of tourism that is now much more in keeping with the area.

A South Downs Authority would be much stronger than the

current Sussex Downs Conservation Board. It would have guaranteed long-term funding from central government and it would also have the statutory right to apply for EEC funding. Unlike the current Board, it would have the power to acquire land or enter into long-term management agreements with land-owners – two essential requisites if conservation projects and recreational activities are to be managed effectively and threatened sites saved. National Park Authorities have had great successes with innovative conservation projects involving farmers and local communities.

Finally, a new South Downs Authority would be managed largely by local elected county, district, parish and town councillors, so giving people a direct say in how their own countryside is looked after.

It is essential that the South Downs are acknowledged as a very special case. The alternative of going for modest legislation to be applied generally is hardly a solution to the problems of our much-loved but vulnerable downland. There is no doubt in my mind that the South Downs should be given National Park status and I hope that the 'first steps' outlined by the government are quickly acted upon and brought to fruition.

I conclude with two quotations which I believe to be apt as an ending to a book such as this. First from Gerard Manley Hopkins:

> What would the world be, once bereft
> Of wet and of wildness? Let them be left,
> O let them be left, wildness and wet;
> Long live the weeds and the wilderness yet.

Finally, a prayer written by the Reverend Hugh Moseley for the Millennium Service at St Simon and St Jude, East Dean, near Eastbourne:

We thank you Lord for the wonder of your creation, for the sea and the sky, the Downs and the Weald, for animal and plant life and for those who have maintained and cultivated this our heritage. Help us truly to appreciate all that You have given us and to conserve its glory for generations to come.

River Meon near Warnford

BIBLIOGRAPHY

AA Book of the Countryside, Drive Publications 1973

Brandon, Peter, *The South Downs*, Phillimore 1998

Bristow, W.S., *The World of Spiders*, Collins 1958

Brown, Leslie, *British Birds of Prey*, Collins 1976

Coulcher, Patrick, *A Natural History of the Cuckmere Valley*, The Book Guild 1998

Coulcher, Patrick, *The Sun Islands*, The Book Guild 1999

Darby, Ben, *The South Downs*, Robert Hale Limited 1979

Ellis, Geoffrey, *The Secret Tunnels of South Heighton*, SB Publications 1996

Gay, Joyce and Peter, *Atlas of Sussex Butterflies*, Butterfly Conservation 1996

Haes, E.C.M., *Natural History of Sussex*, Quadrant House Typesetting, Borough of Brighton, Booth 1977

Hall, P.C., *Sussex Plant Atlas*, Museum of Natural History 1980

Harrison, David, *Along the South Downs*, Butler and Tanner 1958

James, Paul, *Birds of Sussex*, Sussex Ornithological Society 1996

Johnson, Owen, *The Sussex Tree Book*, Pomegranate Press 1998

Lang, David, *Orchids of Britain*, Oxford University Press 1980

Mabey, Richard, *Flora Brittanica*, Sinclair-Stevenson 1996

*Martin, W. Keble, *The Concise British Flora in Colour*, Rainbird 1965

*Mitchell, Alan, and Wilkinson, John, *The Trees of Britain and Northern Europe*, Collins 1982

*Peterson, Roger, Mountford, Guy, Hollom P.A.D., *A Field Guide to the Birds of Britain and Europe*, Collins 1954

*Phillips, Roger, *Grasses, Ferns, Mosses and Lichens of Great Britain and Ireland*, Macmillan 1980

*Phillips, Roger, *Mushrooms and other Fungi of Great Britain & Europe*, New Interintho 1981

Pitt, Frances, *Birds of Britain*, Macmillan 1948

*Readers Digest, *Field Guide to the Butterflies and other Insects of Britain*, Macmillan 1984

Rose, Frances, *The Wild Flower Key British Isles – N.W. Europe*, Frederick Warne 1981

Step, Edward, *Wayside and Woodland Blossoms*, Frederick Warne 1941

*Streeter, David, and Garrard, Ian, *The Wild Flowers of the British Isles*, Midsummer Books 1983

Wickham, Cynthia, *Common Plants as Natural Remedies*, Frederick Muller 1981

Wolley-Dod A.H., *Flora of Sussex*, The Chatford Press 1970

*These are useful books for identification purposes

LATIN NAMES OF FLOWERING PLANTS MENTIONED IN TEXT

Alexanders — Smyrnium olusatrum
Alternate-leaved golden-saxifrage — Chrysosplenium alternifolium
Arrowhead — Sagittaria sagittifolia
Autumn gentian — Gentianella amarella
Autumn lady's tresses — Spiranthes spiralis

Barren strawberry — Potentilla sterilis
Beaked hawk's-beard — Crepis vesicaria
Bee orchid — Ophrys apifera
Bell heather — Erica cinerea
Betony — Stachys offinalis
Bird's-nest orchid — Neottia nidus-avis
Biting stonecrop — Sedum acre
Black horehound — Ballota nigra
Bluebell — Endymion non-scriptus
Blue fleabane — Erigeron acer
Bogbean — Menyanthes trifoliata
Bristly ox-tongue — Picris echioides
Broad-leaved helleborine — Epipactis helleborine
Brooklime — Veronica beccabunga
Bugle — Ajuga reptans
Burnt orchid — Orchis ustulata

Carline thistle — Carlina vulgaris
Charlock — Sinapis arvensis
Chickweed, common — Stellaria media
Chicory — Cichorium intybus
Childing pink — Petrorhagia nanteuillii
Clustered bellflower — Campanula glomerata
Coltsfoot — Tussilago farfara
Columbine — Aquilegia vulgaris
Common bird's-foot trefoil — Lotus corniculatus
Common centaury — Centaurium erythraea
Common comfrey — Symphytum officinale
Common dog violet — Viola riviniana
Common fumitory — Fumaria officinalis
Common gromwell — Lithospermum officinale
Common mallow — Malva sylvestris

Common meadow-rue	Thalictrum flavum
Common poppy	Papaver rhoeas
Common restharrow	Ononis repens
Common rock-rose	Helianthemum nummularium
Common St John's-wort	Hypericum perforatum
Common Sea-lavender	Limonium vulgare
Common spotted-orchid	Dactylorhiza fuchsii
Common stork's-bill	Erodium cicutarium
Common toadflax	Linaria vulgaris
Common twayblade orchid	Listera ovata
Corn mint	Mentha arvensis
Cowbane	Cicuta virosa
Cowslip	Primula veris
Creeping cinquefoil	Potentilla repens
Cross-leaved heath	Erica tetralix
Crosswort	Cruciata ciliata
Curled dock	Rumex crispus
Cut-leaved cranesbill	Geranium dissectum
Dark mullein	Verbascum nigrum
Devil's-bit scabious	Succisa pratensis
Dog's mercury	Mercurialis perennis
Dog rose	Rosa canina
Dove's-foot crane's-bill	Geranium molle
Dropwort	Filipendula vulgaris
Duke of Argyll's teaplant	Lycium barbarum
Dyer's greenweed	Genista tinctoria
Early gentian	Gentianella anglica
Early purple orchid	Orchis mascula
Early spider orchid	Ophrys sphegodes
Early wood violet	Viola reichenbachiana
Elecampane	Inula helenium
Eyebright	Euphrasia nemorosa
Fairy flax	Linum catharticum
Felwort	Gentianella amarella
Field fleawort	Tephroseris integrifolia
Field pansy	Viola arvensis
Field pepperwort	Lepidium campestre
Field scabious	Knautia arvensis
Flowering rush	Butomus umbellatus
Fly honeysuckle	Lonicera xylosteum
Fly orchid	Ophrys insectifera
Foxglove	Digitalis purpurea
Fragrant orchid	Gymnadenia conopsea

Frogbit	Hydrocharis morsus-ranae
Frog orchid	Coeloglossum viride
Garlic mustard	Alliaria petiolata
Germander speedwell	Veronica chamaedrys
Gipsywort	Lycopus europaeus
Gladdon	Iris foetidissima
Glasswort	Salicornia europaea
Golden samphire	Inula crithmoides
Greater knapweed	Centaurea scabiosa
Great mullein	Verbascum thapsus
Greater bladderwort	Utricularia vulgaris
Greater butterfly-orchid	Platanthera chlorantha
Greater sea-spurrey	Spergularia media
Green-flowered helleborine	Epipactis phyllanthes
Hairy violet	Viola hirta
Harebell	Campanula rotundifolia
Heath bedstraw	Galium saxatile
Heather	Calluna vulgaris
Heath speedwell	Veronica officinalis
Heath spotted orchid	Dactylorhiza maculata
Hedge mustard	Sisymbrium officinale
Hedgerow crane's-bill	Geranium pyrenaicum
Hedge woundwort	Stachys sylvatica
Hemp-agrimony	Eupatorium cannabinum
Henbane	Hyoscyamus niger
Herb paris	Paris quadrifolia
Herb robert	Geranium robertianum
Hoary plantain	Plantago media
Honeysuckle	Lonicera periclymenum
Hop trefoil	Trifolium campestre
Horseshoe vetch	Hippocrepis comosa
Hound's-tongue	Cynoglossum officinale
Italian Lords-and-ladies	Arum italicum
Ivy-leaved toadflax	Cymbalaria muralis
Kidney vetch	Anthyllis vulneraria
Lady orchid	Orchis purpurea
Lady's bedstraw	Galium verum
Lady's-smock	Cardamine pratensis
Large flowered evening-primrose	Oenothera glazioviana
Lesser butterfly-orchid	Platanthera bifolia
Lesser celandine	Ranunculus ficaria

Lesser sea-spurrey	Spergularia marina
Longleaf	Falcaria vulgaris
Lords-and-ladies	Arum maculatum
Lousewort	Pedicularis sylvatica
Man orchid	Aceras anthropophorum
Marjoram	Origanum vulgare
Marsh cinquefoil	Pontentilla palustris
Marsh marigold	Caltha palustris
Meadow clary	Salvia pratense
Milk vetch	Astragalus glycyphyllos
Moon carrot	Seseli libanotis
Moschatel	Adoxa moschatellina
Moth mullein	Verbascum blattaria
Mugwort	Artemisia vulgaris
Musk orchid	Herminium monorchis
Narrow-leaved helleborine	Cephalanthera longifolia
Nettle-leaved bellflower	Campanula trachelium
New Zealand pygmy weed	Crassula helmsii
Nottingham catchfly	Silene nutans
Opposite-leaved golden-saxifrage	Chrysosplenium oppositifolium
Orange mullein	Verbascum phlomoides
Oxeye daisy	Leucanthemum vulgare
Oxford ragwort	Senecio squalidus
Pheasant's-eye	Adonis annua
Ploughman's spikenard	Inula conyzae
Prickly sow-thistle	Sonchus asper
Primrose	Primula vulgaris
Purple loosestrife	Lythrum salicaria
Pyramidal orchid	Anacamptis pyramidalis
Ragged robin	Lychnis flos-cuculi
Ragwort	Senecio jacobaea
Raspberry	Rubus idaeus
Red bartsia	Odontites verna
Red clover	Trifolium pratense
Red star-thistle	Centaurea calcitrapa
Red helleborine	Cephalanthera rubra
Red valerian	Centranthus ruber
Rock samphire	Crithmum maritimum
Rosebay willowherb	Chamerion angustifolium
Rottingdean sea-lavender	Limonium hyblaeum
Rough marsh mallow	Althaea hirsuta
Round-headed rampion	Phyteuma orbicularae

Rue-leaved saxifrage	Saxifraga tridactylites
Sainfoin	Onobrychis viciifolia
Salad burnet	Sanguisorba minor
Sanicle	Sanicula europaea
Scarlet pimpernel	Anagallis arvensis
Scentless mayweed	Tripleurospermum inodorum
Sea campion	Silene uniflora
Sea-heath	Frankenia laevis
Sea kale	Crambe maritima
Sea pink	Armeria maritima
Sea radish	Raphanus raphanistrum subspecies maritimus
Silverweed	Potentilla anserina
Small-flowered buttercup	Ranunculus parviflorus
Small hare's-ear	Bupleurum baldense
Small-leaved sweetbriar	Rosa agrestis
Small scabious	Scabiosa columbaria
Small toadflax	Chaenorhinum minus
Snowdrop	Galanthus nivalis
Soapwort	Saponaria officinalis
Solomon's-seal	Polygonatum multiflorum
Spotted medick	Medicago arabica
Spiny restharrow	Ononis spinosa
Spurge laurel	Daphne laureola
Squinancywort	Asperula cynanchica
Starry clover	Trifolium stellatum
Stemless thistle	Cirsium acaule
Stinking hellebore	Helleborus foetidus
Tall melilot	Melilotus altissimus
Tamarisk	Tamarix gallica
Thrift	Armeria martima
Toothwort	Lathraea squamaria
Tormentil	Potentilla erecta
Traveller's-joy	Clematis vitalba
Tuberous pea	Lathyrus tuberosus
Twayblade	Listera ovata
Viper's bugloss	Echium vulgare
Wallflower	Erysimum cheiri
Water soldier	Stratiotes aloides
White bryony	Bryonia dioica
White helleborine	Cephalanthera damasonium
White horehound	Marrubium vulgare
Wild basil	Clinopodium vulgare

Wild clary	Salvia horminoides
Wild clematis	Clematis vitalba
Wild daffodil	Narcissus pseudonarcissus
Wild liquorice	Astralagus glycyphyllos
Wild mignonette	Reseda lutea
Wild radish	Raphanus raphanistrum subspecies raphanistrum
Wild rasberry	Rubus idaeus
Wild strawberry	Fragaria vesca
Wild thyme	Thymus drucei
Wood anemone	Anemone nemorosa
Wood spurge	Euphorbia amygdaloides
Wood-sage	Teucrium scorodonia
Woody nightshade	Solanum dulcamara
Yarrow	Achillea millefolium
Yellow archangel	Lamiastrum galeobdolon
Yellow bird's-nest	Monotropa hypopitys
Yellow horned-poppy	Glaucium flavum
Yellow pimpernel	Lysimachia nemorum
Yellow rattle	Rhinanthus minor
Yellow toadflax	Linaria vulgaris
Yellow-wort	Blackstonia perfoliata

INDEX

Bold numbers refer to plate numbers. Main reference pages and pages showing sketches are in bold italics.

246

INDEX